SOCIOBIOLOGY: SENSE OR NONSENSE?

EPISTEME

A SERIES IN THE FOUNDATIONAL,

METHODOLOGICAL, PHILOSOPHICAL, PSYCHOLOGICAL,

SOCIOLOGICAL AND POLITICAL ASPECTS

OF THE SCIENCES, PURE AND APPLIED

Editor: MARIO BUNGE

Foundations and Philosophy of Science Unit, McGill University

Advisory Editorial Board:

VOLUME 8

MICHAEL RUSE

University of Guelph, Guelph, Ontario, Canada

SOCIOBIOLOGY: SENSE OR NONSENSE?

D. REIDEL PUBLISHING COMPANY

DORDRECHT : HOLLAND / BOSTON : U.S.A.

LONDON : ENGLAND

Library of Congress Cataloging in Publication Data

CIP

Ruse, Michael.
    Sociobiology, sense or nonsense?

    (Episteme ; v. 8)
    Bibliography: p.
    Includes index.
    1.   Social behavior in animals.   2.   Wilson, Edward Osborne,
1929-   —Sociobiology.   I.   Title.
QL775.R87   591.5   78–21000
ISBN 90-277-0943-2
ISBN 90-277-0940-8   pbk.   (Pallas edition)

_____

Published by D. Reidel Publishing Company,
P.O. Box 17, Dordrecht, Holland

Sold and distributed in the U.S.A., Canada, and Mexico
by D. Reidel Publishing Company, Inc.
Lincoln Building, 160 Old Derby Street, Hingham,
Mass. 02043, U.S.A.

For my wife April

TABLE OF CONTENTS

ACKNOWLEDGEMENTS

Many people kindly sent me preprints of their writings on sociobiology. Others listened patiently to my half-formed ideas. Thanks should go to: Peter Achinstein; Richard Alexander; Mario Bunge; Brian Calvert; William Durham; William Hughes; David Hull; Scott Kleiner; Noretta Koertge; Richard C. Lewontin; Kevin Mutchler; Fred Suppe; Edward O. Wilson. In the circumstances, it is particularly important to emphasize that no one but myself is responsible for any part of this book. As always, Judy Martin did an exemplary job with the typing.

I am obliged to the following for permission to reproduce illustrations:

Figure 2.1 is redrawn from *Darwin's Finches: An Essay on the General Biological Theory of Evolution* by D. Lack. Copyright © 1947 by Cambridge University Press.

Figure 2.2 is reproduced from *The Philosophy of Biology* by M. Ruse. Copyright © 1973 by Hutchinson Publishing Group Limited.

Figure 3.1 is reproduced from *Entomophagus Insects* by C. P. Clausen. Copyright © 1940 by McGraw-Hill Book Company.

Figures 3.2, 3.3. and 4.1 are redrawn from *Sociobiology and Behavior* by D. Barash. Copyright © 1977 by Elsevier Publishing Company.

Figure 3.2 originally appeared Copyright © 1972 by the Aldine Publishing Company. Reprinted with permission, from *Sexual Selection and the Descent of Man* (New York: Aldine Publishing Company.)

Figure 3.3 originally appeared in *American Zoology* 14 (1974).

Figure 4.1 originally appeared in *Behavioural Science* 20 (1975).

Figures 8.1 and 8.2 are reprinted from *Human Ecology* 4 (1976). Copyright © the Plenum Publishing Corporation.

Figures 8.3 and 8.4 are reprinted from *The Quarterly Review of Biology* 51 (1976).

*"Descended from monkeys?
My dear let us hope that
it is not true! But if it
is true, let us hope that
it not become widely known!"*

*(The wife of the Bishop of
Worcester upon hearing of
Charles Darwin's theory of
evolution through natural
selection.)*

CHAPTER 1

INTRODUCTION

In June 1975, the distinguished Harvard entomologist Edward O. Wilson published a truly huge book entitled, *Sociobiology: The New Synthesis*. In this book, drawing on both fact and theory, Wilson tried to present a comprehensive overview of the rapidly growing subject of 'sociobiology', the study of the biological nature and foundations of animal behaviour, more precisely animal *social* behaviour. Although, as the title rather implies, Wilson was more surveying and synthesising than developing new material, he compensated by giving the most thorough and inclusive treatment possible, beginning in the animal world with the most simple of forms, and progressing via insects, lower invertebrates, mammals and primates, right up to and including our own species, *Homo sapiens*.

Initial reaction to the book was very favourable, but before the year was out it came under withering attack from a group of radical scientists in the Boston area, who styled themselves 'The Science for the People Sociobiology Study Group'. Criticism, of course, is what every academic gets (and needs!); but, for two reasons, this attack was particularly unpleasant. First, not only were Wilson's ideas attacked, but he himself was smeared by being linked with the most reactionary of political thinkers, including the Nazis. Second, although some of the members of the critical group were Wilson's colleagues — indeed, hitherto had been considered friends — the attack was made publicly (in the form of a letter to the *New York Review of Books*, following a sympathetic review by the geneticist C. H. Waddington) and without the courtesy of prior warning to Wilson.

As can be imagined, attack was followed by counter-attack, and the bitterness escalated. As also did the circle of interest, professional and public, until finally the dispute was accorded that ultimate American accolade, cover-story treatment by *Time* magazine! Certainly, for all of his troubles, Wilson can feel satisfied that he has helped raise public consciousness about sociobiology, although he must also take comfort in the fact that general sentiment has been one of sympathy towards him for the way in which he was persecuted. And indeed, some of Wilson's initial attackers have regretted the way in which he was criticized, even though they may still endorse the essential content of the attack.[1]

1

Now that tempers have cooled somewhat and we are starting to move away from the time of the most harsh salvos, it would seem that we might more profitably hope to look at the sociobiology controversy: not at the personalities particularly, but rather at the various ideas being expressed. Certainly, the whole question of the true nature and basis of animal social behaviour seems worth studying. And if we include the nature and basis of human social behaviour, then the interest and importance of the inquiry seems much magnified. Moreover, whatever we may feel about the particular actions and motives of the various disputants in the sociobiology controversy, they certainly seem to have earned the right as scientists to have their ideas taken seriously. As intimated, prior to the publication of *Sociobiology* Wilson was rightly considered one of the world's leading insect biologists. And on the other side, the company is, if possible, even more prestigious, for we find amongst the Boston critics (to name but two) the brilliant population geneticists, Richard Lewontin and Richard Levins. Lewontin, in particular, has given modern population biology a major forward impetus, because of the way in which he has brought molecular theory and findings to bear on traditional problems.[2]

Therefore, because the subject seems important and because the disputants seem to be the sort of people who would have something worth saying, I want to consider, in this book, the sociobiology controversy. As stated, my concern is not with personalities or even particularly with motives. I want to see what case can be made for sociobiology and what case can be made against it. Because my inquiry is intended to be fairly abstract and far reaching, I shall not restrict my exegesis of sociobiological claims exclusively to Wilson's writings, but shall feel free to refer generally to the work of sociobiologists (as, of course, Wilson himself does). Conversely, although I shall obviously be referring in some detail to the objections of the 'Science for the People' critics, I shall feel free to cast my net more widely there too.

As I begin I should perhaps enter a personal note, not so much by way of apology but more by way of explanation. I am trained and work as a philosopher of science, not as a biologist. It might therefore seem somewhat impertinent of me even to try to write such a book as this: the sociobiology controversy is a biological controversy and ought therefore be handled by biologists. However, I think I can legitimately and appropriately enter the fray. Thomas Kuhn in his stimulating work, *The Structure of Scientific Revolutions*, has pointed out that when one gets major scientific conflicts and disagreements one finds frequently that crucial differences rest not so much on matters of pure science (whatever that might mean), but more on matters which for want of a better word we might call 'philosophical'. The differences involve logic, methodology, metaphysics, and so on. I do not know how widely Kuhn's general analysis of science holds: certainly, I shall be giving reasons for showing how the sociobiology controversy causes difficulties for this analysis; but in this question of philosophy I think, in this instance, Kuhn is right.[3] As we shall see, much of the sociobiological controversy goes beyond science to matters philosophical: at least, to matters that philosophers talk about at great length! For this reason, I presume to write on the sociobiology controversy. Of course, if I get my biology wrong I expect to be criticized by biologists; but that is nothing to what I expect from philosophers if I get my philosophy wrong. I should add that my arrogance equals my presumption for I hope that what I have to say will be of interest both to biologists and philosophers. It is, for this reason, that I try always to provide elementary biological and philosophical background. I know that philosophers need the biology, and I suspect that equally the philosophy will be of value to biologists.

The structure of this book is as follows: First, following the introduction of some essential biological theory, I shall present the major theoretical and factual claims made for non-human sociobiology. Since I am not writing a popular introduction to sociobiology, I shall not feel pressed to mention absolutely everything. I hope indeed to cover enough that a reader new to sociobiology can get a fair idea of the subject; but I shall be writing always thinking of the objections levelled against sociobiology. Second, I shall repeat my presentation but dealing with the claims made for human sociobiology. I should add now that I shall not be dealing with earlier popular writers about supposed biological bases of human behaviour, for example Robert Ardrey and Desmond Morris. For reasons that will be explained, people like Wilson feel that they have given human sociobiology a whole new approach, and because I tend to agree with them and because the earlier writers have not been involved in the recent controversy, I shall ignore these writers in this book.

Third, I shall turn to the various criticisms that have been made of sociobiology (non-human and human). Because both sides will now have been presented, I shall evaluate the merits of the criticisms as we go along. Fourth, I shall ask what, if anything, might be the long-term scientific implications of sociobiology. In particular, I shall look in detail at some recent speculations by Wilson about the possible future effects of sociobiology on the social sciences. Fifth, and finally, I shall ask what, if anything, might be the long-term philosophical implications of sociobiology. In particular, I shall look in detail at some speculations by Wilson about the possible effects of sociobiology on philosophy.

NOTES TO CHAPTER 1

[1] Short histories of the sociobiology controversy can be found in Wade (1976) and Currier (1976). The Boston critics' first attack on Wilson was Allen *et al.* (1975). They followed with an expanded version in *BioScience*, Allen *et al.* (1976), and in an even-more expanded version Allen *et al.* (1977). Wilson replied publicly to these critics in Wilson (1975c) and Wilson (1976).

[2] Wilson collaborated with the late Robert MacArthur on a key work in ecology, MacArthur and Wilson (1976). His own major work on insects is Wilson (1971), and he has just published, in collaboration with George Oster, what may prove to be a fundamental contribution to the theory of insect behaviour, Oster and Wilson (1978). Lewontin's major work is on the genetic variation within groups, Lewontin (1974). I discuss his ideas in Ruse (1976a) and Ruse (1977a). Levins is best known for his work in theoretical biology, Levins (1968).

[3] As I shall suggest later, the sociobiology controversy is strongly reminiscent of the controversy following the publication of Charles Darwin's *Origin of Species*. There also, philosophical matters were important, although there are also many difficulties for a Kuhnian analysis. See Ruse (1970), (1975a), (1978); Hull (1973), (1978a).

CHAPTER 2

THE BIOLOGICAL BACKGROUND

In this chapter, I want to introduce some fairly basic biological ideas and theory, so that these can then be presupposed for the rest of the book. Obviously, I do not want to introduce the whole of biology, but rather those aspects which have some bearing on sociobiology. Therefore, the guiding thread at this point will be the nature of sociobiology and the way in which it is supposed to relate to the rest of biology. Possibly, some readers interested primarily or exclusively in human behaviour might regret the fullness of my treatment, and they may be tempted to skip ahead. I think this would be a mistake. Perhaps one thing, more than anything, distinguishes both the claims and the style of sociobiologists from previous writers about the biological bases of human social behaviour, namely the way in which the sociobiologists believe that they are the first to approach human behaviour backed by a solid foundation of tested biological theory. Of course, we may conclude later that the links the sociobiologists see both between their work on social behaviour in the non-human world and the rest of biology and between their work in the non-human world and social behaviour in the human world are nothing like as tight as they themselves suppose; but these are things that will have to be investigated, not assumed at the outset. For this reason, consequently, if only out of fairness to the sociobiologists, it is important to establish as solid a biological background as is possible. Let us therefore turn to Wilson's definition of sociobiology and work backwards to general biological principles.

2.1. SOCIOBIOLOGY AS BIOLOGY

At the beginning of *Sociobiology: The New Synthesis*, Wilson writes: "Sociobiology is defined as the systematic study of the biological basis of all social behavior". (Wilson, 1975a, p.4.) We are therefore interested in animal behaviour, or, more precisely, animal behaviour inasmuch as it involves interaction with other animals. We are not directly concerned with most of the morphological and other features of organisms, for example the thick coat of the polar bear to keep out the cold, although, of course, our concern does extend to non-behavioural features which do in some sense get involved in social behaviour — weapons for fighting, and so on. (Note that here, as always,

5

the word 'social' is used in a broad sense which covers many kinds of interaction with other animals, including some behaviour which, in another sense, we might label positively anti-social. We shall return later to the meaning of 'social'.)

But what is the 'biological basis' to which Wilson is referring in the above definition? The answer will come quickly to anyone who has read Charles Darwin's classic *On the Origin of Species*, for in that work not only does Darwin attempt to explain the physical features of organisms by means of his theory of evolution through natural selection, but also he applies his theory to behavioural features of organisms, for example the slave-making behaviour of various kinds of ants. For Darwin, and for biologists after him, the key to the biological understanding of animal features, and we include here animal behavioural features, is evolution through natural selection, And, in this respect, we find that Wilson and his fellow sociobiologists stand firmly and openly in the Darwinian camp. They want to understand animal social behaviour as a product of Darwinian evolution. "Darwin's theory of evolution through natural selection is central to the study of social behavior . . ." (Trivers, 1976, p.v. See also, Dawkins, 1976, p.1; Wilson, 1975a, p.4.)

Now, we must tread carefully here or we run a danger of misunderstanding the sociobiologists' intentions. First, as might well be expected, since Darwin's major impetus evolutionary thinking has taken great strides. Obviously, although the sociobiologists think of themselves as 'Darwinian', and although, as we shall see later, there is a strong continuity between the past and the present, the sociobiologists draw their inspiration and guidance essentially from the modern theory of evolution: the theory usually called the 'synthetic' theory of evolution. Second, connected with this first point, we must understand what precisely might be meant or implied in taking a modern evolutionary approach to a problem. For most non-biologists, evolution is usually identified with the fossil record, and that is that. This being so, an evolutionary approach to animal behaviour might seem frustrating to say the least. Almost by definition, behaviour is one thing that does not get fossilized, and so it might seem that, at best, one is going to spend one's time grasping at straws, trying to infer hypothetical behaviour from those characteristics of animals that do get fossilized. However, whilst this kind of inference does have its place in the evolutionary-studies spectrum, to a biologist there is, in fact, very much more involved in the taking of an evolutionary approach to a problem. The approach involves not just the fact and path of evolution, but also the mechanism of evolution. Then, in turn, the approach can make reference to all the disparate areas of biological inquiry united by the mechanism: morphology, systematics, embryology, biogeography, and so on. Consequently, inasmuch as they want to take an evolutionary approach, it would seem that sociobiologists want to integrate studies of animal social behaviour, with this family of theories (or sub-theories). This being so, let us therefore take a closer look at modern evolutionary theory. Then, towards the end of the chapter, we can work our way back to sociobiology. (See Ruse, 1969a, 1972, 1973a, 1973a for a fuller treatment of the following points.)

2.2. PRINCIPLES OF GENETICS

Paradoxically, although the modern theory of evolution can properly be said to be founded on Darwin's ideas, probably the best expository place to begin is with the major non-Darwinian component: the *gene* concept. It is almost a truism to point out that different organisms have different characteristics — some are tall, some are short; some are heavy, some are light; some eat meat, some eat plants; some breed prolifically, some breed virtually or absolutely not at all; and so on. The biological causal factors behind all of these characteristics, the units of function, are the genes. They are carried inside the nuclei of the cells of the body, on strips called the *chromosomes*. Every cell has genes, and although different organisms have different sets of genes, within the main body every cell has the same set. And it is these genes, today identified at the molecular level as deoxyribonucleic acid (DNA), which interact ultimately to cause bodily characteristics.[1] (See George, 1964; Strickberger, 1968; Watson, 1970.)

As one rapidly finds out, if one did not already know, nothing in biology is very straightforward. There are some complications and qualifications that must be made to the picture of the gene. The most important point to be made is that, generally speaking, genes in animals do not occur singly, but in pairs. More precisely, the chromosomes in cells can be, and at certain times are, paired. Every gene has a mate on the paired chromosome, the position on the chromosome being known as the *locus*. Genes tend to come in different forms, causing differences in characteristics, but normally only genes from one set can occupy a particular locus: the members of such a set being called *alleles*. From what has been said, it follows that in a particular organism at a particular locus (that is, on the two corresponding loci on the paired chromosomes), there will be two alleles, which might be the same or different. If the alleles are the same, then with respect to that locus the organism is said to be *homozygous* (it is a homozygote); if the alleles are different, the organism is *heterozygous* (heterozygote). An organism might be homozygous with respect to one locus, but heterozygous with respect to another.

Given the pairing of alleles, there are various ways in which they might and do express themselves at the physical level (known as the *phenotypic* level, as opposed to the level of the genes, the *genotypic* level). In particular, in a heterozygote it might be that the effect of one allele swamps entirely the effect of the other allele (that is, phenotypically, it is as if the organism were homozygous for the first allele). In such a case, the first gene is said to be *dominant*, and the second to be *recessive*. An allele might be dominant over a second allele, but recessive to a third. Sometimes, there is a fairly tight isomorphism between a gene and a particular phenotypic effect. Sometimes, however, a gene affects more than one characteristic. Such genes are said to be *pleiotropic*. Conversely, when more than one gene (that is, alleles from different loci) is involved in forming a characteristic, the genes are said to be *polygenes*.

Finally, considering the gene from the viewpoint of function, we must make reference to a point to which we shall be returning: it is most important to note that, strictly speaking, it is not quite proper to say without qualification that genes *cause* phenotypic characteristics. Apart from anything else, in certain circumstances in loose parlance one wants to say that it is not the genes which cause characteristics, but the environment. The reason why my chairman has a healthy tan and I am pallid white is not so much a function of our differing genes, but that he spent his vacation in Spain whilst I was holed up in my office working. However, strictly speaking, it is not really satisfactory either to say that we have a simple dichotomy of genetically caused characteristics and environmentally caused characteristics. In reality, it is always the genes in conjunction and interaction with the environment which cause such characteristics. Consider, say, something like animal height. It may well be, indeed certainly will be, that if we keep the environment absolutely constant, there are sets of alleles which can cause fluctuations in height. Some genes make an animal taller; other genes make an animal shorter. In this sense, height is genetic. On the other hand, it is almost equally certain that we can get height fluctuations whilst holding the alleles constant and varying the environment. In this sense, height is non-genetic or environmental. In a way, therefore, talk of things being genetic or non-genetic, that is talk of things being genetically caused or environmentally caused, is probably rather misleading, because one can feel fairly certain that nature is sufficiently ingenious to stretch apart the causal components of a great many characteristics in this way. And, this is not to mention what can happen when humans take a hand. (Hull, 1978b.)

However, although it is not strictly true to say simply that genes cause characteristics, or that the environment causes characteristics, it would be too extreme to drop entirely the genetic-environmental causal dichotomy. Indeed, in a way, we shall learn that it is because sense can be made of it and because this sense is so important that the sociobiological controversy has been able to occur. Some characteristics are going to develop into what they are, more or less no matter what the environment that occurs naturally; on the other hand, some characteristics are going to be sensitive to almost any environmental change. Take just one case as a paradigm for both kinds of characteristic: bird songs. (Hinde, 1970.) There are clearly both genetic and environmental components to bird songs: no genes, no songs; no food, no songs. However, some kinds of birds are almost totally insensitive to the environment when it comes to their songs. They will sing the same song as their fellows even if they are brought up in total isolation. For other birds, however, the environment has a crucial effect on their songs. Brought up in isolation they do not sing, and brought up around other kinds of singing birds they copy them. Clearly there is an important difference here and having due regard to the various points just made, we can speak of the former kind of song as being 'genetic' and the latter as being 'environmental'. In the case of behaviour like singing, we can also speak of 'instinctive' as opposed to 'learned' behaviour; and, as is well known, sometimes the distinction is spoken of being one of 'nature' as opposed to 'nurture'.

Returning now to the gene itself, let us look next at the other side of the gene, namely the gene as the unit of inheritance. It is the genes, or copies of them, that are passed on from one generation to the next, *via* the sex-cells. Unlike the normal ('somatic') cells, the sex-cells contain only one set of un-paired genes, one such gene coming from each locus. In sexual organisms, normally an organism receives one such unpaired set (a *haploid* set) from each parent, and thus the initial cell again contains pairs of genes (is *diploid*). According to rules discovered by the father of the modern theory of inheritance, the European monk Gregor Mendel, considering any locus it is purely a matter of chance which allele of a parental pair gets passed on to the offspring, and what happens at one locus does not affect the chances of what happens at other loci. Of course, in speaking of 'chance' one does not imply that ultimately things are uncaused, and in fact it has been discovered that in certain circumstances what happens at one locus can affect other loci.[2]

2.3. POPULATION GENETICS

So far, we have been considering matters just at the level of the individual.

The genetics of the individual can readily be generalized to deal with genes, their distribution, and their inheritance in populations. The basic extension, the foundation of 'population genetics', is called the Hardy—Weinberg law, after its two co-discoverers. Elementary symbols help at this point. Suppose we have a population of organisms, effectively infinite in size, that they inter-breed at random, that at some particular locus there are two possible alleles, A_1 and A_2, and that the proportions are p:q. The law then states that for all succeeding generations, the ratio will stay at p:q and that, moreover, no matter what the initial distributions, unless there are disrupting factors, the ratios of the various genotypes will be as follows:

$$p^2 A_1 A_1 + 2pq A_1 A_2 + q^2 A_2 A_2,$$

($A_1 A_1$ is a homozygote for A_1, and so on).

The Hardy—Weinberg law functions in population genetics rather like Newton's first law of motion functions in Newtonian mechanics: it provides a background of stability, in effect telling one that if nothing happens then nothing happens. Like mechanics, where population genetics takes flight is in considering precisely what disruptive factors there might be and how they might work: by 'work' in this context being meant how disruptive factors might affect gene ratios from one generation to the next. Two major potential factors are *mutation* and *selection*. (See Li, 1955; Mettler and Gregg, 1969.)

First, something might go wrong in the copying process as new genes are made, and thus a new form of gene, a 'mutant', appears, and this, in turn, affects the phenotype. Mutation is random in the sense that it does not occur in response to an organism's needs, indeed most is deleterious to its possessor, but on average it can be quantified. Obviously, in the long run, mutation is the raw stuff of evolution, for without it forms would never change. However, on its own, mutation is not going to take us very far very fast. And on its own it is not going to explain what is probably the most significant feature of the organic world, certainly that feature which differentiates it most from the inorganic world, its *adaptedness*: the fact that organisms are not just random things but seem as if they were designed, with their characteristics being *adaptations* helping their possessors to survive and reproduce. (See Ayala, 1970; Mayr, 1974; Hull, 1973, 1974.)

And so this brings us to the second potential factor disrupting or altering gene ratios, the Darwinian contribution to population genetics, *natural selec-tion*. Organisms are born, live, and die. Their genes are not going to be repre-sented in succeeding generations unless they reproduce, and, moreover, their genes are not going to be as well represented in future generations unless the organisms reproduce at least as well as their parents. But there is no automatic guarantee of reproduction. An organism might get killed before it reaches its reproductive prime, say by a falling rock. Of course, probably this would be just a random accident, but there are other problems of a more constant and repeating nature — most particularly, there is the fact that there are other organisms all striving to survive and reproduce. The conflict with others could well cut down an organism's survival and reproduction.

Now, what is argued is that certain genes, or alleles, give their possessors characteristics which make for just that little bit more efficiency in the struggle to survive and reproduce: they give their possessors the adaptive edge. Thus these 'fitter' genes have that little better chance than the genes of competitors in getting through in increased numbers to the next generation. (Considering just ratios in a large population, it is not necessary that every possessor of a certain kind of gene always triumph over the possessor of another kind of gene, but that on average that they do.) Drawing on the analogy of a human picking one kind of organism to breed rather than another, biologists speak of the more favourable kind of gene, that is the fitter gene, being 'selected' rather than the other (or having a higher coefficient of selection than the other). Clearly, selection understood in this sense can have a direct effect on gene ratios, and as we shall see it is believed that, ultimately, evolution is the effect of selection sifting through constantly appearing new mutations. Moreover, through this process, adaptation is explained: it is the cumulative effect of the success of ancestors. Fairly obviously, this notion of adaptation is a relativized notion. In an important sense, there is no ultimate objective adaptation, as was rather supposed when adaptation was explained through God's creative design. To the Darwinian biologist, adaptation is all a matter of what works in a particular situation, and may well be something else in another situation. (Ruse, 1977a.)

Selection, like mutation, can be quantified, and through this population genetics has been worked into a sophisticated mathematical theory. There is no need for us here to enter into details; but, with an eye to future discussion, there are a couple of aspects to the concept of selection which merit some comment. These concern the ways in which selection might at times act to maintain gene ratios as they stand, and the level or levels at which selection can act.

2.4. SELECTION AS PRESERVER OF THE STATUS QUO

Although evolutionists believe that significant change of gene ratios is brought about and directed by selection, it is not difficult to show that, under certain circumstances, selection can hold things (*i.e.* gene ratios) fairly stable in a society. One such circumstance is if there is a selective advantage within the population for rareness. This is not an outlandish possibility, for it could well be that one has a population subject to predation and that the predators have to learn to recognize their prey. Clearly, any oddness of form in the population would help the possessor because the predators would be less likely to know it. Selection would therefore favour the odd form. But obviously, before very long, the odd form would no longer be so very rare! Hence, predators would start to recognize the form and selection would be stepped up against it. Eventually, what one would expect would be all the various forms held in a kind of balance within the population, with their various ratios being a function of (amongst other things) the ease with which the predator recognizes the forms as potential prey. In other words, the fairly stable gene ratios would be a function of the different selective forces acting on them. (Sheppard, 1975.)

Another circumstance leading to this kind of balance of gene ratios, one much discussed in the literature, revolves around so-called 'superior heterozygote fitness'. If, given two alleles at a locus, the heterozygote can be shown to be at a selective advantage to, or fitter than, either homozygote, then it is easy to show that so long as the situation prevails, the alleles will be held indefinitely at a balance within the population. Intuitively, one can see this because the heterozygote on average will always contribute to the next generation. Consequently, one will always get both alleles in the next generation, and the balance will come about because the various selective forces will eventually cancel each other out. The classic case of such balanced heterozygote fitness involves genes which cause so-called 'sickle-cell anaemia'. Although homozygotes for the sickle-cell gene usually die in childhood of anaemia, the genes are held stably in certain human African populations because the heterozygote for the gene have a natural immunity to malaria and are thus fitter than homozygotes without the gene. (Raper, 1960; Livingston, 1967; 1971; Dobzhansky *et al.*, 1977.)

It is worth noting a couple of consequences of these kinds of balanced situations just discussed. First, they mean that within populations there is no genetic (or phenotypic) uniformity. Different kinds of alleles are always present. Hence, should the environment or something else change, thus setting up radically new selective pressures, because there is already much genetic variation, fairly rapid genetic change (*i.e.* change of gene ratios) can occur at once. There is no need to wait exclusively for new mutations: Second, these situations point dramatically to a fact raised above, namely that there is no question of absolute adaptive advantage or fitness. In the scarcity case, nothing is totally better than anything else, or totally worse. In the superior heterozygote fitness case, it could well be that although an allele works marvellously with some other allele, when paired with a twin it is absolutely dreadful, even killing the possessor before reproduction. Clearly, one consequence is that in a population with such a phenomenon, one is going to have a constant supply of phenotypes much less fit than (that is, at a reproductive disadvantage to) other phenotypes within the population.

There is, one should add, some debate about how common these various balancing mechanisms actually are in nature; but thanks to brilliant pioneering work by Lewontin we now know that populations do contain a great deal of genetic variation, and undoubtedly some of this does express itself at the phenotypic level, although there is still much question about how much and about what kinds of selective effects it has. (Lewontin, 1974; Ayala et al., 1974.)

2.5. THE LEVEL OF SELECTION

Turning now to our second point, an important question which must be asked about natural selection is precisely who benefits by it? It may seem that the answer is so obvious that the question is not really worth asking. Without a doubt, the individual organism is going to benefit, and through this, ultimately, the organism's species. But is this so? Are the interests of an individual organism and of its group, particularly the species, always identical? Or, to put the matter another way, could one have some characteristic which was of value to the individual but not to the group, and, if so, could selection favour such a characteristic; conversely, could one have some characteristic of value to the group but not the individual, and could selection favour it? (Lewontin, 1970.)

Clearly, often the interests of an individual and of a group coincide. An individual has a characteristic; it helps the individual survive and reproduce; and inasmuch as the individual is a member of a particular species, the species is helped to survive and reproduce. But, prima facie it seems equally clear that a characteristic might help an individual but not the group, and vice versa. Suppose, for instance, a particular area can support only a certain number of a certain organism. It is in the species' interests that one gets as close to that number as possible. The individual, however, wants to maximize its own total number of offspring. (I am using anthropomorphic language here and do not intend to imply conscious intention.) Thus, the interests of individual and group can conflict. A certain characteristic might make an individual double its offspring number. This could take the population number above its critical point, thus causing a fairly drastic decline, way below the possible maximum.[3] But still, overall, the individual is ahead, even accepting its share of the decline, because it has more offspring than it would have without the characteristic in question. Conversely, a characteristic might help the group but not the individual. Apparently, a self-sacrificing altruism, where an organism heroically gives up its life for its fellows, will help the species but seemingly not the individual organism.

The question of whether one can properly interpret characteristics as of value to the individual but not (or only incidentally to) the group, or to the group but not (or only incidentally to) the individual, has troubled and divided biologists for many years. As also has the related question of whether one can properly distinguish an 'individual selection', promoting characteristics of value to the individual, from a 'group selection', promoting characteristics of value to the group (and if one can, whether both exist). Indeed, the co-discoverers of natural selection, Charles Darwin and Alfred Russel Wallace, differed on these matters. They both recognized that certain hybrid organisms have certain characteristics that render them sterile, and they both recognized that such characteristics could be of value to the parent species because they prevent future maladaptive hybrid offspring. However, whereas Darwin thought that the sterilizing characteristics have to be accidental because selection could never fashion something harmful to the individual however much it may help (unrelated) conspecifics, Wallace thought that precisely because of their value to the parent species such characteristics could be formed through selection. (Both Darwin and Wallace allowed that aversion to breeding with other species' members could be produced by selection, because this could be of value to the individual. (Darwin and Seward, 1903, 1, pp.287–99.))

The debate continued until the publication, in 1962, of V. C. Wynne-Edwards's *Animal Dispersion in Relation to Social Behaviour* brought matters to a head. Wynne-Edwards argued cogently and at length for the efficacy of group selection, and for his efforts he was rewarded with the dubious honour of having stirred to action a host of biologists, most notably G. C. Williams, determined to prove him wrong. (Williams, 1966.) And in fact, as matters stand today, the general concensus is that, with the possible exception of a few very specialized cases, Wynne-Edwards was indeed wrong. Selection is almost always for the individual and not for the group, or more precisely for the group only if it is for the individual. In other words, inasmuch as a characteristic is to be explained in terms of selection, it is to be explained in terms of its promoting an individual's genes rather than the collective genes (the 'gene-pool') of the group. (Wilson, 1975a.)

Naturally, this denial of the existence of group selection creates something of a dilemma for biologists. Many characteristics do seem as if they were created by group selection, for they seem of value primarily or exclusively to the group not the individual. However, biologists do not throw up their hands in despair, denying that such characteristics have any selective values or causes. Rather, biologists strive to give alternative explanations based on individual selection. We shall be meeting many of these kinds of explanation as this book progresses, for it is particularly with respect to animal behaviour that group selection explanations seem most plausible. But perhaps here it will help illustrate the problem and the attempted resolution, if an example is given.

A fact well-documented by ornithologists is that many birds have clutch-sizes that are constant within the species: for example, petrels lay one egg, pigeons two, gulls three, and plovers four. Moreover, the non-accidentality of these numbers is underlined by the fact that if eggs are removed, then the birds lay more to bring the number back to original. To a group selectionist, the correct explanatory argument is that in not exceeding the specific clutch-size the birds are practicing a kind of birth-control for the good of the group. If reproduction went completely unrestrained, then before long the whole group would stand in danger of starving to death. Hence, argues the group selectionist, selection acts at the level of the group, in the sense that it wipes out any individual which has a tendency to have a clutch-size not in the group's reproductive interests. For the individual selectionist, however, the birth control is for the good of the individual parents. As argued by the late David Lack, if a bird lays less eggs then it is at an obvious reproductive disadvantage, but if it lays more eggs then it is also at a disadvantage, because, given the burden of extra offspring to rear, the probabilities of raising successfully even the original lesser number of offspring decline. (Lack, 1954, 1966.) Hence, argues the individual selectionist like Lack, selection acts at the level of the individual, in that it wipes out any individual which has a tendency to have a clutch-size unlike the specific norm, for such an individual is less fit than its fellows.

By this point, one must be wondering why modern biologists have this obsession with the individual. Why are they so biased against the group, and so keen to explain in terms of the individual's advantage? The answer is simply because it seems that group selection will not work. As Darwin pointed out, however valuable something may be for the group, if it is not also of value, directly or indirectly, to the individual, it just will not get passed on. Selection works first on the individual, and if a characteristic is to be preserved, it must make its case there. "Natural Selection cannot effect what is not good for the individual . . ." (Darwin and Seward, 1903, 1, p. 294.) Suppose, to take the example of the plover, that although a clutch-size of four is of value to the group, the individual could just as easily raise five. Any genes which favoured the larger number would at once be promoted, because selection starts with the individual. Say one had a population where all but one or two raised four (or whatever proportion actually reach maturity) and that the exceptions raised five (or the appropriate proportion). In the next generation, the genes of the exceptions would be better represented, and so on, even though the eventual result might be the total collapse of the whole population. In short, reason biologists, one just must seek out advantage to the individual for everything, whatever the appearances, because it is through the individual that selection begins.

2.6. THE THEORY OF EVOLUTION

Having now attended to these important side points, let us turn back to our main theme. So far, we have been talking about population biology or genetics. Where does this all tie in with the theory of evolution? Evolution seems to be something which requires vast periods of time, and we have been considering genetic changes at the level of individual generations — certainly not really huge numbers of generations. However, there is nothing paradoxical here. A major tenet of modern evolutionary thought is that what happens on the large scale, involving great quantities of time, is no more than a summation of events happening on the small scale. In other words, the modern theory of population genetics — involving, as it does, gene transmission, mutation, selection, and the like — provides the mechanism of evolution. Then, armed with this mechanism, evolutionists can turn to particular areas of biological inquiry which are of interest to them. (For more on evolutionary theory, see Simpson, 1953; Mayr, 1963; Dobzhansky, 1970; and Maynard Smith, 1975.)

Thus, for instance, let us for a moment consider the area concerned with organic geographical distributions, biogeography, and take but one problem of great concern to them: the distinctive flora and fauna on oceanic islands. For example, on the Galapagos Archipelago, to highlight something of crucial importance in turning Darwin towards evolutionism and, since, brilliantly analysed by Lack (1947), we find the most peculiar distribution of a sub-family of birds, (*Geospizinae*, known as 'Darwin's finches'). Although all the birds are fundamentally very similar, they fall into four genera and fourteen species, and moreover they are distributed irregularly throughout the islands of the archipelago, some species being found only on one island and others on several. (See Figure 2.1.)

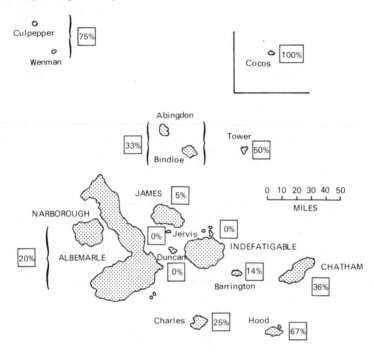

Fig. 2.1. (From Lack, 1947.) Percentage of endemic forms of Darwin's finches on each island, showing effect of isolation.

To explain this phenomenon, evolutionists suppose that ancestral finches flew out to the islands from the South American mainland (where finches can still be found), and they argue that because of the isolating barriers caused by the seas between islands, different finches will have evolved apart, taking their descendents away from the descendents of common species' members. And, to justify their arguments, evolutionists appeal to their knowledge of population genetics, believing that it is this which provides them with the precise mechanics of how such evolution occurs. The evolution of the finches required generation after generation of changes of gene ratios guided by forces encaptured in the principles of population genetics: mutation, selection, more mutation, more selection, and so on. Moreover, it is suggested that because populations contain so much genetic variation, any sub-set isolated on an island, cannot be typical of the parent body, because there are no typical members. Hence, this difference in itself will cause fairly rapid evolution away from the parent body, because with its limited variation the sub-set will respond differently to selective forces from the way that the original larger group responds. (This is Ernst Mayr's 'founder principle'.)

In short, population genetics provides the background foundation for arguments in geographical distribution (or biogeography), and it functions in the same way for other areas of evolutionary biology: morphology, systematics, paleontology, and so on. Of course, each discipline has its own peculiar claims — for example, in biogeography one has claims pertaining to the actual way or ways in which organisms can, or cannot, get transported around the globe — and there is a certain amount of borrowing as premises of conclusions from other disciplines. The paleontologist, in particular, has to assume as given many claims which other evolutionists can infer from the study of still-living organisms. But, essentially, population genetics yields the unifying core between the various sub-areas of evolutionary biology. In a way, therefore, one can conveniently think of evolutionary theory as having a kind of fan-like form with population genetics at the apex, as illustrated in Figure 2.2. (Ruse, 1972, 1973. See also Hull, 1974.)

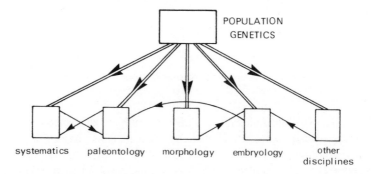

Fig. 2.2. (From Ruse, 1973.) (In this figure, the rectangles represent various disciplines; the double lines the links between population genetics and other areas — such links are actually supposed to exist; and the single lines links between the subsidiary disciplines — although such links do exist, those shown in the figure are just illustrative, they do not necessarily denote particular instances.)

2.7. SOCIOBIOLOGY AS PART OF EVOLUTIONARY THEORY

And so this now brings us right to the heart of what the sociobiologists aim to do. Basically and simply, they want to add on sociobiology as another member of the evolutionary family. They want to fit in the study of animal social behaviour as another sub-discipline linked by the common core of population genetical biology. Or, to be more precise, they want to develop the theory of animal social behaviour, showing its genetic background, thus re-emphasizing the theory's proper place in the evolutionary spectrum. And indeed, Wilson is quite explicit on this matter, for at the beginning of *Sociobiology* he states these intentions openly: "This book makes an attempt to codify sociobiology into a branch of evolutionary biology and particularly of modern population biology." (Wilson, 1975a, p.4.)

Because the intentions of the sociobiologists are so straightforward, no more need be said by way of scientific preliminary. We are now in a position to turn on their actual speculations and findings, which we shall, in fact, do almost at once. But, precisely because the sociobiologists' intentions are so straightforward, we can make a couple of philosophical points which might help to clarify subsequent discussion. Therefore, these points will be raised briefly in bringing this chapter to an end.

First, note that whatever innovations they may introduce, in a very important sense sociobiologists want to tread down a path beaten out by others. They want to be orthodox evolutionists. In fact, as we have seen, they stand in line with Darwin who, like them, also wanted to explain animal social behaviour in terms of evolution powered by natural selection. But, for the moment just staying with the present and with modern population genetics which incorporates Darwinian selection, we see that acceptance of this is the starting point for sociobiologists. They do not seek to overthrow the theory and findings of population genetics. Of course, the sociobiologists want to extend and stretch theory as they tackle their own particular problems, but this is no more than is done by any evolutionist. In this sense, therefore, to use Thomas Kuhn's well-known language, the sociobiologists are "normal" scientists working within their "paradigm": they are doing "research firmly based upon one or more past scientific achievements, achievements that some particular scientific community acknowledges for a time as supplying the foundation for its further practice". (Kuhn, 1970, p.10.)[4] In particular, sociobiologists intend to do research firmly based upon past scientific achievements in population genetical biology.

One should add, however, that the neatness with which one can fit Kuhn's analysis at this point is a little self-defeating. In the last chapter, I pointed out that in the sociobiological controversy we seem to have the kind of appeal to philosophy that Kuhn finds at times of crucial stress and change in science: scientific revolutions. Unfortunately, Kuhn argues also that these times are when scientists switch allegiances, from one theoretical framework or paradigm to another. Hence, we might have expected a paradigm switch here, something which does not seem to be occurring. Rather, what we are getting is the extension of an already-established paradigm to an area where other paradigms have been proposed. (Barash, 1977.) Thankfully, Kuhn's problems are not our problems, and so we do not have to try to resolve the paradox. As everyone else seems to have done, we can use Kuhn's insights without having to bother about the truth of his overall thesis. Sociologically and psychologically, we seem to have all the marks of a Kuhnian-type scientific revolution, but, epistemologically, we do not.

Of course, an obvious comment here is that although the sociobiologists may claim to be using existent theory, in fact they are not. As Kuhn points out, often in paradigm switches one gets a carry-over of language, thus concealing the radicalness of the revolution. Perhaps sociobiologists claim to be using normal population genetics, but surreptitiously — intentionally or otherwise — have substituted their own theory. This means that we do indeed have an epistemological Kuhnian revolution. Clearly, however, we cannot address ourselves to this objection until we have seen more of the sociobiologists' work, so let us reserve judgement on this until later. The stated intention of sociobiologists is to work from an already-accepted body of theory.

The second point of philosophical interest arising out of the sociobiologists' intentions concerns the overall structure of evolutionary theory, and thus indirectly sociobiology. There is debate today between philosophers of science as to the proper interpretation and analysis of scientific theories. Some argue that scientific theories are properly seen as 'hypothetico-deductive' systems, where conclusions can be seen to follow deductively from the initial axioms, the ultimate hypotheses of the theories. Other philosophers deny this argument. Fortunately, this is another philosophical dispute whose ultimate resolution is not vital to our purposes; but it does have a bearing on our study in this wise. Both those for, and those against, the hypothetico-deductive thesis agree that the case for it can be made most strongly in the physical sciences: from whence it came in the first place. As it stands, evolutionary theory is not very hypothetico-deductive. Frequently, to say the least, there are big gaps between premises and conclusions, with the possible links being sketched in, hinted at, or hypothesized. Although there are some compact bodies of deductive theory in evolutionary studies, for instance the population genetical core, frequently in evolutionary studies when one speaks of things 'following from' or 'throwing light on' one has weaker connections than deduction in mind.

As might be expected, a little absence of firm evidence has never deterred philosophers, and those for the hypothetico-deductive thesis argue that all of this merely proves that evolutionary theory as it stands is a sketch of a full theory. The opponents obviously draw different conclusions. But, whatever the correct answer may be on this point, the dispute does highlight the 'looseness' of evolutionary studies. Inasmuch, therefore, as sociobiology is part of the evolutionary family, we ought not judge it by standards more strict than we would apply to the rest of evolutionary theory. (For more on the structure of evolutionary theory, see Ruse, 1973a, 1977a; Williams, 1970; Goudge, 1961; Hull, 1977.)

Enough now of the biological background to sociobiology. Let us turn next to sociobiology in action.

NOTES TO CHAPTER 2

[1] In this brief discussion, I am ruthlessly ignoring irrelevancies, like those microorganisms with RNA as the ultimate carrier of heredity.

[2] Genes on the same chromosome are 'linked' and with respect to each other follow different rules of inheritance.

[3] This is the sort of thing that happens when a population strips its area of vegetation, thus causing wholesale starvation and disease.

[4] A paradigm for Kuhn is a kind of scientific world-view or *Weltanschaung*, into which most scientists are locked most of the time. Scientists working within a paradigm, doing 'normal' science as opposed to revolutionary science when they challenge a paradigm, accept as inviolable the basic premises of the paradigm. (Suppe, 1974.)

CHAPTER 3

THE SOCIOBIOLOGY OF ANIMALS

The concern in this chapter is with the sociobiological fact and theory of animal behaviour, more specifically, animal social behaviour. I shall take in turn the subjects of aggression, sexuality, parenthood, and altruism. Although this will not exhaust the sociobiological *corpus*, it will be fully representative. Inevitably, there will be some overlap, because the topics chosen are by no means fully separate.

3.1. AGGRESSION: THE ETHOLOGICAL VIEWPOINT

The whole question of animal aggression has fascinated evolutionists right back to Darwin: indeed, it fascinated pre-evolutionary biologists no less, as they wrestled to harmonize nature "red in tooth and claw" with the supposedly benevolent omnipotence of the Christian God. (Ruse, 1975b, 1977b.) For Darwin, who used the struggle for existence to fuel natural selection, aggression was a fundamental, vital, and pervasive facet of animal existence, although as a matter of fact his 'struggle' covered a far wider range of things than just two animals battling to the death; extending metaphorically for instance to a cactus 'battling' against drought, and a pretty flower 'struggling' with its fellows to get attention from insects. (Ruse, 1971b.)

In recent, years, the whole question of animal aggression has been much illuminated, not to say popularized, by the writings of a number of so-called ethologists, most particularly Konrad Lorenz. In his fascinating and deservedly well-known book, *On Aggression*, Lorenz argues at length that the traditional view of animal aggression as an inevitable bloody battle to the end is quite mistaken, particularly as applied to conflict between animals of the same species. Certainly, one gets a struggle to the death when, say, a lion attacks an antelope: if one did not, then the lion would never get its supper. But fights between animals of the same species, a very common kind of animal aggression, are quite otherwise. They involve a kind of social interaction, something we might be loathe to apply to a prey—predator situation. The fighting is always restrained, involving ritual, bluff, and violence of a non-fatal kind. And, moreover, there are appeasement gestures that can be made by an animal losing a conflict, so that the winner will not follow through to the kill. Dogs, for instance, will present their bellies to an overwhelming attacker, at once defusing the fury of this aggressor. Furthermore, it seems that, generally, animals do not like the taste of their conspecifics; consequently, animals that prey on others have no motive to prey on their fellow species' members. Indeed, argues Lorenz, so effective are the restraining mechanisms, that: "Though occasionally, in territorial or rival fights, by some mishap a horn may penetrate an eye, or a tooth an artery, we have never found that the aim of aggression was the extermination of fellow-members of the species concerned." (Lorenz, 1966, p.38.)

So much for fact; what about the theory? The aggression of an animal against another to defend its nest or to gain its supper is easily explicable in terms of Darwinian selection. But why should there be any aggression at all between animals of the same species, although there undoubtedly is, and why then given this aggression should there be so much restraint? At this point, Lorenz invokes group selection hypotheses: aggression between conspecifics exists to pick out the best members of the species, so that these will provide the breeding stock for the future, the species' best interests being served by having the best members as parents. Equally, however, it is in the species' best interests not to have any of its members wiped out, particularly since the weaker usually include the younger, and so group selection perfects all of the limiting mechanisms. "The environment is divided between the members of the species in such a way that, within the potentialities offered, everyone can exist. The best father, the best mother, are chosen for the benefit of the progeny. The children are protected." (Lorenz, 1966, p.38.)

Complementing this view of the animal world are a few wise but sad words about the human predicament. Somehow, in our case, selection has gone wrong. We no longer seem to have the near-infallible restraining mechanisms of the brutes. When faced with fellow humans, we are killers! An 'evil intraspecific selection' set in during the early Stone Age, and as a consequence we humans can no longer keep our animosities within check. We therefore wage wholesale war on other humans. Science confirms religion, for, verily, humans are tainted with original sin.

This scenario of animal aggression, taken almost as gospel by many these days, has been challenged by sociobiologists both with respect to fact and to theory. Starting first with fact, let us turn to Wilson's synthesis.

3.2. WHAT IS ANIMAL AGGRESSION REALLY LIKE?

The important and interesting questions concern aggression between animals of the same species. No one denies that, generally speaking, aggression between

members of different species involves fairly straightforward Darwinian
functions of food, defence, and so forth. Now considering aggression within
species, there is agreement with Lorenz and those that argue with him that
widespread aggression exists, that much of it is genetic or innate, and that
indeed much of it is restrained. Very often, animals do not escalate their fighting into all-out warfare, particularly not at first. However, there is strong disagreement with Lorenz as to the supposedly almost-universally and invariably
limited nature of aggression between members of the same species.

Beginning with the insect world, Wilson is able to give a long list of species
where fights to the death occur between fellow species' members, even being
accompanied by cannibalism. Indeed, in certain parasitic species of Hymenoptera (this is the insect order which includes wasps, bees, and ants), the larvae,
for a while, transform into a strange form that seems specifically adapted to
kill and consume all other conspecific larvae that are occupying the same
insect host! Lethal conflict occurs until only one is left, and then the larvae
transform on again. (Wilson, 1975a, p.246.) See Figure 3.1.

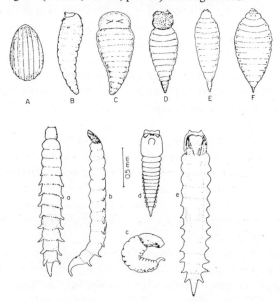

Fig. 3.1. These two series show the successive developments of the larvae of two species
of parasitic wasp (upper *Poecilogonalos thwaitesii*, lower *Collyria calcitrator*). They show
graphically how one stage (upper D, lower c–e) has adaptations specifically for killing
the larva's conspecifics. (From *Entomophagous Insects* by C. P. Clausen. Copyright ©
1940 by McGraw-Hill Book Company. Used with permission.)

Moving up the scale, lest it be thought that only in the invertebrate world do murder and cannibalism obtain, we find that in fact they are widespread in the animal kingdom, right up to and including the higher vertebrates. For all of their restraint, many animals kill their fellows, and they are often not beyond eating them. Indeed, far from popular opinion being true, the human species comes out looking rather well. Thus, lions sometimes kill each other, and, given the chance, the males are not beyond eating the cubs fathered by others. Hyenas have a murder rate above Detroit's, and they too on occasion make meals of each other. In the primate world, murder is not unknown either. For instance, in India, langurs live in packs with females dominated by one male. If another male can break in and take over, it tries to kill all of the young. Even in chimpanzees, murder and cannibalism sometimes occurs. And switching across to the bird order, we find that there too sometimes there is struggle to the death. (Wilson, 1975a, Chapter 11, esp. pp.246–7.)

By this point, one may be wondering why Lorenz was so absolutely wrong. The sociobiologists argue that the mistakes are both possible and actual, because to establish the full truth about animal aggression (and indeed about animal behaviour in general), one must have very long-time studies of animal behaviour in the wild, and only now are these beginning to obtain. About murderous behaviour in animals, Wilson writes: "I have been impressed by how often such behavior becomes apparent only when the observation time devoted to a species passes the thousand-hour mark." (Wilson, 1975a, p.247.) And somewhat to underline his point, Wilson goes on to add that one murder per thousand hours is a great deal of violence by human standards, and in fact he suggests that with present information, even taking account of human warfare, compared to the rest of animal creation human beings are starting to look like very peaceful beings indeed. We are far less dangerously aggressive than many animals, not excluding apes.

The ethological claims as to the facts of animal aggression are therefore challenged. As might be expected, Lorenz's group selection hypotheses are also questioned. In particular, the sociobiologists want to work from, and only from, individual selection. Now, in a sense, they can do this easily; perhaps even more easily than someone like Lorenz. The sociobiologists make no *a priori* assumptions about the good of the species, and hence have no need of special explanations as to why one organism might attack a fellow. Thus, all other things being equal, in the eyes of the sociobiologist the parasitic wasp larva is indifferent as to whether it is attacking a fellow or a member of a different species. Another organism means food, or competition, or something. And more generally, the sociobiologists, Wilson in particular, see animal aggression as explicable in terms of a competition "for a common resource or requirement that is actually or potentially limiting". (Wilson, 1975a, p.243.) There is only so much to go round, and aggression ensures that an animal gets its share, or more. Since conspecifics usually want the same thing, there is no wonder that there is competition and aggression within species. (See Figure 3.1.)

Furthermore, points out Wilson, aggression can vary according to need. In particular, when resources are very limited, aggression often escalates, or animals show other sorts of bizarre behaviour. Cats which are overcrowded become despotic and frenzied attacks occur on some which become pariahs. Rats show hypersexuality, homosexuality, cannibalism and other behaviour which is, in the circumstances, 'abnormal', (Wilson's word). All of this can be directly understood in terms of Darwinian adaptive advantage.

With the explanation of aggression in terms of competition for resources we are already in the domain of theory. However, as sociobiologists, we have still not explained the most surprising fact about animal aggression, namely that despite the limitations that we may want to put on Lorenz's account, animals do nevertheless show a great deal of restraint in their conflicts with fellows. How can this be explained, together with the fact that sometimes the conflict escalates? More particularly, how can this be explained in terms of individual selection rather than the group selection which is an anathema to sociobiologists? As Wilson points out, the answer must be that unrestrained aggression is more costly to the individual than restrained aggression. However, although he himself offers some suggestive proposals, a far more complete analysis is at present being formulated in Britain. Since it shows well the strengths, but also some of the current limitations of sociobiology, it merits a brief look.

3.3. EVOLUTIONARY STABLE STRATEGIES

The approach to be discussed derives from the branch of applied mathematics known as game theory, and it owes its genesis to John Maynard Smith and a number of co-workers. Central to Smith's work is the notion of an 'evolutionary stable strategy', an ESS for short, which he characterizes as a strategy such that "there is no 'mutant' strategy that would give higher reproductive fitness". (Maynard Smith, 1972, p.15.) What is meant by this is a situation where one has a population with a number of possible forms, and where given the particular ratio of forms that actually obtains, individual selection does not favour one form over any other. In short, the population is balanced or stable, for one would not expect one form to be increased at the expense of others. Perhaps this notion of an ESS can best be illustrated by giving a simple model or two, specifically those which have been proposed to show how limited aggression with the possibility of all-out violence can be maintained in populations. (See also Maynard Smith 1972, 1974, 1976; Maynard Smith and Price, 1973.)

Consider first a population of organisms, with two possible kinds: Hawk and Dove, or Mouse, (these are just names for the types). When Hawks meet a fellow-species' member they fight in an all-out fashion until they win or are seriously injured. Doves fight in a ritualized manner, until they or their opponent become bored and drop out; they always retreat before real aggression. By setting up the right numerical values, one can show that neither total Dove nor total Hawk is an ESS in a population, but that a certain proportion of each is. That is to say, one can show that a population of Doves is not stable, for individual selection would favour a mutant Hawk. But then neither is a population of Hawks stable, for individual selection at this point would favour a mutant Dove! However, at a certain ratio of Hawks to Doves, an individual would not be any better off being a Dove rather than a Hawk, or *vice versa*. Individual selection would therefore hold the polymorphic population stable.

Quantifying everything with almost the gay abandon of Jeremy Bentham,[1] suppose that a win is worth +50 points, a loss 0 points, serious injury or death −100 points, and wasted time −10 points. When two Doves meet, we know that someone will win (+50) and that a lot of time will be lost (−10 points each). On average therefore, a Dove can expect: $(50 - 10 \times 2) \times \frac{1}{2} = +15$, out of a Dove–Dove contest. Similarly, a Hawk meeting Hawk can expect, assuming that the battle is swift and is bloody for the loser: $(50 - 100) \times \frac{1}{2} = -25$. And when Hawk meets Dove, since the Hawk always wins at once, for the Hawk we have +50 and for the Dove 0. We can see that a population of Doves is not pursuing an ESS, because a mutant Hawk would start to spread (advantage +50 over +15 from any encounter with another population member, which latter is by stipulation a Dove). On the other hand, although an individual Hawk will always beat an individual Dove, a population of Hawks is not pursuing an ESS, because a mutant Dove would start to spread (advantage 0 over −25). In fact, although one might fear that a population such as this would oscillate wildly between the extremes of total Dove and total Hawk, there is indeed an ESS, namely when the ratio of Doves to Hawks is 5:7. At this point, a mutant from Hawk to Dove or *vice-versa* would be no better off, because the average pay-off to both Hawk and Dove is $6\frac{1}{4}$. (One can calculate this by reckoning the probability that a Hawk would meet a Dove, and so forth.) In other words, what this model shows is that we could have a population, continuing indefinitely in a stable fashion, where (as in nature) one gets some ritualized non-harmful aggression, and (as in nature) some very real and dangerous aggression. And it is all caused and maintained by individual selection.

Before going on to anything more sophisticated, a couple of points should be made. First, if it be thought closer to real life to have the same organism sometimes showing Hawkish behaviour and sometimes Dovish behaviour, the model can be interpreted as showing that a 5:7 Dove–Hawk behaviour ratio in every individual is an ESS. Second, most interestingly (although perhaps not unexpectedly) the model yields different results from a group selection model. Under the above individual selection model, one expects the population to evolve towards the ESS, with an average pay-off of $6\frac{1}{4}$. Under a group selection model, one expects the population to evolve towards that which is best for the population. Obviously, a population of total Doves is better than a 5:7 Dove–Hawk split, because then the average pay-off is 15! No one in such a population gets seriously injured; at worst, they just waste time. Of course, as we saw in the last section, the trouble is that, in real life, animals do get injured and killed in intra-specific combat; so for this reason if for no other the individual selection model is more promising.

What we have presented so far is a very simple model, with but two kinds of sharply defined behaviour. However, one can readily extend the notion of an ESS to cover more complex situations, which are probably closer reflections of reality. For instance, Maynard Smith and Price (1973) have devised a more subtle model where one has five types: Hawk; Dove (which they call 'Mouse'); Bully who comes on as a Hawk, but quickly changes to a Dove or flees if the opponent is also Hawkish; Retaliator who comes on as a Dove, but who turns Hawkish if the opponent is Hawkish; and Prober-Retaliator, who plays Retaliator most of the time, but every now and then tests out the mettle of the opponent by turning Hawkish. By assigning certain plausible values the authors show that probably the population would evolve to mainly Retaliator and Prober-Retaliator, with just a small number of Mice. In fact, Prober-Retaliator only really gains on Retaliator when there are Mice (who will never fight back); but as the authors point out, in any real population, one is almost bound to get some Mice: the young, the old, the sick, and so on.

This game-theoretic analysis can be extended in other ways too. For example, suppose that instead of actually fighting, organisms engage in a 'war of attrition', trying to outlast the opponent, through staring or threatening or the like. Here, the loss is not physical injury but valuable time. It can readily be shown that what selection will favour is not an animal staring or threatening for a fixed time, but varying the period — the exact range depending on various factors, like the prize and the cost and the like. Selection will also favour the development of real 'poker faces' — if unpredictability is the key, it would be stupid to let one's opponent know what one's own state of mind is. Also let us mention that, with suitable modifications, the analysis can cover 'asymmetrical' contests, where one of the partners in the contest is assumed to have a certain built-in advantage or disadvantage, over and above the actual abilities involved in the contest itself. Here, outside factors can really make a difference as to which of alternative ESS's actually evolve. (Maynard Smith and Parker, 1976.)

Enough has now been said to give the reader a proper flavour of the game-theoretic approach to animal aggressive behaviour. What can be said in conclusion about this approach? Without, at this point, probing too deeply, there seem to be two obvious comments about its present degree of success.

3.4. STRENGTHS AND LIMITATIONS OF THE GAME-THEORETIC APPROACH

First, in favour of the game-theoretic approach, as has already been intimated strongly, the models based on the approach do seem closer to what ever-increasing evidence is coming to show is the true reality about animal intra-specific aggressive encounters. Quite apart from the difficulties with group selection outlined in the last chapter, such a selection just does not adequately explain why it is that intra-specific aggression sometimes escalates to the limit. On the other hand, Maynard Smith's models do show how restraint can be maintained and also they show what does seem to be a fact, that sometimes real violence erupts. If an organism is not prepared to fight things out on occasion, then its supposed readiness to be Hawk-like is seen to be pure bluff. Moreover, at this point there seems to be physiological evidence for the game-theoretic approach (particularly the model which sees a preponderance of Retaliator types), for experimental evidence shows that when in pain (*e.g.* as when attacked), many animals do not flee but respond with much increased aggression. (Maynard Smith and Price, 1973, p.18.)

The extension of the analysis to non-violent wars of attrition also finds support. Thus, to take the example of the Siamese fighting fish, *Betta splendens*, Maynard Smith writes:

Ritual conflicts between males are usually followed by escalated fights, in which one or both rivals may be seriously injured. Conflicts between females however often end (typically after 5–15 minutes) with the surrender of one fish, without escalated fighting. Simpson followed such conflicts in detail, measuring the frequency and timing of particular components of the ritual. He found no significant difference between the frequencies with which eventual winners and eventual losers performed particular acts, except during the last 2 minutes of a contest, when the eventual winner could be recognized from the fact that her gill covers were erected for a larger proportion of the time. The fact that the winner could not be distinguished from the loser until close to the end of a contest fits well with the prediction from game theory. (Maynard Smith, 1972, p.24.)

Finally, whilst considering support for the game-theoretic approach, we have the matter of asymmetrical contests. An obvious application here is to the much-discussed phenomenon of territoritality. It is well known that in many species of animal, individuals mark out their own particular territories, and, what is most interesting, other members of the species tend to respect these territories — even though the territories may be of value to them. Tinbergen, for example, showed that male sticklebacks will defend their own territories, but that if artificially introduced into the territories of others they will attempt to flee. (Tinbergen, 1971.) The game-theoretic approach can explain such phenomena, as soon as one assumes that conflicts are dangerous or time-wasting or the like. An ESS is: "If at home, stay; and if at someone else's home, don't waste time, get out." Note, incidentally how, at this level, an equally effective ESS is: "If at home, get out; and if at someone else's home, stay"! However, probably there are other reasons why squatting does not evolve as a general policy, even though there is some evidence that it does sometimes. For instance, the resident will usually actually know the territory, and thus in this respect be at an actual physical advantage over the intruder.

Nevertheless, having now said what is in favour of a game-theoretic approach to animal aggression, we come to the second obvious comment to be made about it. At the present, we have what seems a very promising approach, but hardly what one could call a well-established theory. The models so far devised give results which correspond roughly to what happens in nature, but to date there is hardly any question of rigorous testing. As is more than obvious, the game theorists choose values which make their results come out to what seems to happen in nature; but in themselves the various values chosen are pretty arbitrary. There is no actual check on what the values might be in real life, nor of assessment of predictions against observed behaviour. Is there indeed a species where winning outright is worth +50 points, whereas being killed is a 100 point penalty, and does this species really have a stable ratio of Hawks to Doves, specifically 7:5?

In short, we ought modestly to conclude that much of the work at the moment shows that the models are plausible, rather than that they absolutely must be accepted.

3.5. SEX AND SEXUAL SELECTION

So far we have been considering animals engaged in various forms of conflict. However, if their genes are to be sent on to future generations, the organisms themselves must at some point stop fighting and start reproducing! Since many animals are sexual, that is to say there are two kinds and one of each kind is required to make an offspring, this raises the whole question of the behaviour which surrounds reproduction, something which has been of intense interest to sociobiologists.

In a sense, of course, one has the prior question of why there should be sex at all. As most of us know only too well, the course of true love tends not to run that smoothly, and it might therefore seem that an asexual organism — one that can reproduce without the help of another — would be at a selective advantage. The advantage to sex seems to lie in the fact that new mutations, causing favourable characteristics, can, through sex, be brought together that much more rapidly. Suppose organism x has mutation a and organism y has mutation b. Even though ab might make a terrific combination, without sex one cannot bring the existing mutations together, but must wait until x mutates to b or y to a. However, whilst there is obviously some truth in this answer, it cannot be denied that it runs dangerously close to being a group selective answer. It may well pay the species to have an ab combination, but does the hope that an offspring of x will pick up a b compensate for the effort of having to find a mate? Various models have been proposed to answer this and like questions, although it is probably true to say that the matter is still in a state of flux without firm answers. (Maynard Smith, 1975, pp.185–91; Williams, 1975.)

But, let us assume that we have sex. Right from the start, it has been recognized that sex has interesting and important evolutionary implications. In particular, in the *Origin of Species* Charles Darwin introduced an entire evolutionary mechanism that was centred on sex. Darwin's main mechanism was natural selection, which was a function of such things as the need to find food and shelter; but one of his secondary mechanisms was *sexual selection*, which was a function of the struggle to find mates. Darwin probably arrived at his notion of natural selection by analogy from artificial selection, and it was probable also that it was this analogy which led him to sexual selection:

at least, which led him to believe that sexual selection was sufficiently important to be promoted as an evolutionary mechanism in its own right. (Ghiselin,
1969.)

Breeders, when selecting, tend to do so for one of two reasons. Either they
want profit, as when the farmer wants shaggier sheep or heavier cows, or they
want pleasure, as when the fancier wants prettier pigeons. The first of these
aims (and successes) led Darwin to natural selection and the second to sexual
selection, or, more precisely, to the two varieties of sexual selection which
he recognised: male combat and female choice. Drawing on the analogy of
breeders selecting for fighting dogs and cocks, Darwin argued that sometimes
males fight between themselves for females, and thus one gets the evolution
in males of offensive weapons, directed at other male members of their respective species. Drawing on the analogy of breeders selecting for beautiful
animals, Darwin argued that sometimes females choose from among the
males, and thus one gets the evolution in males of beautiful characteristics,
directed at the females of their respective species. (Darwin, 1859, pp.87–90.)

As soon as it was proposed publicly, sexual selection faced controversy,
and this has continued down through the years. (Vorzimmer, 1970; Campbell,
1972.) The widespread existence of sexual dimorphism (i.e. differences between the sexes) is indisputable, and given the bizarre forms that it can take
at times — as in the peacock — it is equally indisputable to a Darwinian evolutionist that some form of selection plays a major role in fashioning it. Moreover, generally, no one has seemed to want to say that Darwin was completely
wide of the mark. Male combat does certainly exist and males seem to have
weapons to aid them in it: one thinks, for example, of the huge elephant seals
who battle brutally for the possession of harems. And it cannot be denied
that the peacock does have a magnificent display of tail-feathers. However,
many people have felt that female choice smacks unduly of unwarranted
anthropomorphism, imputing to the peahen aesthetic qualities peculiar to man,
and they have tried to eliminate it. Thus, Alfred Russel Wallace explained
such dimorphisms not in terms of the male being beautiful, but rather in
terms of the female being shabby! He argued that there is a selective premium
on unobtrusive females, since they, as main carers for their young, need
camouflaging protection from predators. (Wallace, 1870, pp.231–61.) And
others have drawn attention to the fact that although sexual selection rests on
a struggle for reproduction rather than all-out survival, so also does natural
selection at times, and hence they have tended to demote sexual selection and
include its insights into natural selection. (Lack, 1966.)

The sociobiologists have rather revived sexual selection — as one might have expected, since it rests very much on evolutionary implications of animal behaviour, and that after all is their field! Nevertheless, as might also be expected, although in many respects they seem closer to Darwin than most of his successors, particularly with respect to such matters as female choice, the sociobiologists do not present things in quite Darwin's terms. Perhaps their greatest advance centres on a notion developed by Robert L. Trivers, namely that of 'parental investment', and it will therefore be appropriate and convenient to structure our discussion around this notion. (Trivers, 1972.)

3.6. PARENTAL INVESTMENT

In order to introduce this notion, we should begin at the beginning. Suppose one has the rudimentary startings of sexuality, with the two sexes newly formed. Each puts forward an equal sex-cell (a 'gamete'). Obviously (*i.e.* it seems reasonable to suppose!) were a parent to produce a somewhat larger cell, with a bit of its own food supply, it would be at a selective advantage over its fellows. However, immediately then there would be a selective advantage to another organism producing less costly (and therefore probably more) smaller gametes, which upon fusing with the larger gametes could share in the extra food supply. Consequently, one would get the evolution of two types of sex-cell: large cells, produced in relatively small numbers, and small cells, produced in relatively large numbers. Whether by definition or as a matter of fact, 'femaleness' implies producing large cells and 'maleness' implies producing small cells.

Now enters the notion of parental investment. Both parents, male and female, want to produce offspring.[2] However, someone has to bring up baby. If either parent can slough off the work on the other, then so much the better (from an evolutionary viewpoint) because the parent is then free to go around looking for another mate, capable of producing yet more offspring. But, of course, the other parent wants to do just the same, and so the question now arises: who has more to lose? Fairly obviously, it is normally the female who is caught in this dilemma: she is the one who has 'invested' more, or, more pertinently, she is the one who is going to have to make most effort if she decides to cut her losses and start again. Hence, straightaway we see that we have a conflict of interests. The male wants to fertilize and get away and go on to the next; the female wants to be fertilized, but she wants to hang on to the male for help or otherwise get compensation. Thus we get different selective forces, and generally speaking what we ought to find (and what we do in fact find) is that males are more interested in fertilizing many females, and females are more interested in rearing those offspring that they do have. It should be added that there are exceptions to this picture and we shall have to consider these shortly.

Somewhat more formally, we can state matters thus: Parental investment is defined as "any investment by the parent in an individual offspring that increases the offspring's chance of surviving (and hence reproductive success) at the cost of the parent's ability to invest in other offspring". A large parental investment is one which greatly decreases the ability of the parent to invest in other offspring. Any individual has a total possible parental investment and thus having regard to the total possible offspring that an individual could produce, one can work out a kind of average parental investment per off-spring by an individual. Considering the two sexes in a species, they do not necessarily have the same average parental investment per offspring; however, each sex can only produce the same total number of offspring as the other sex. Consequently, the sex having the greater average parental investment than the other becomes a limiting resource for that other with the lesser in-vestment. And this leads the way to various selective pressures acting on the individuals in a species. (See Figure 3.2.)

Now, how might these various selective pressures take place and what form might they take? First, most obviously we are going to get competition be-tween the members of the sex which invests less to breed with the members of the sex which invests more. Any organism of the lesser-investment sex whose genes promote characteristics which aid that organism to breed more successfully than its fellows is going to pass on those very genes in increased proportions to the next generation. Normally, from what has just been pointed out above about the relative sizes (and thus investments) in the sex-cells, it is the females who are the limiting resource for the males, and thus there is competition and resulting selective pressures on the males. This is, of course, Darwin's sexual selection through male combat, and it is most widely documented empirically throughout the animal world. Males of many, many species battle for the females, and they have all sorts of adaptations to help them to battle successfully. (Wilson, 1975a, pp.318–35.)

It must be pointed out, however, that the theory of parental investment by no means necessarily implies that the females of a species will necessarily be the sex of limiting resource. Parental investment refers to all the effort an organism puts into its offspring. This includes such things as nest-building, care after birth, and so forth. Because of the differences between the gametes and the associated physiological differences (like the female mammal carrying the embryo) this usually means that the female is indeed the sex of limiting

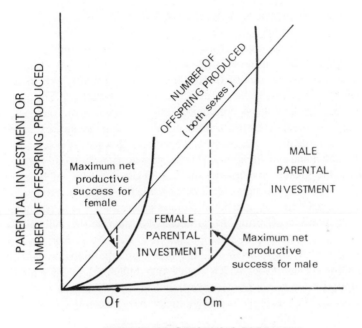

PARENTAL INVESTMENT OR NUMBER OF OFFSPRING PRODUCED

NUMBER OF OFFSPRING PRODUCED (both sexes)

MALE PARENTAL INVESTMENT

Maximum net productive success for female

FEMALE PARENTAL INVESTMENT

Maximum net productive success for male

O_f O_m

NUMBER OF OFFSPRING PRODUCED

Fig. 3.2. (From Trivers 1972, via Barash 1977.) This graph compares parental investment and reproductive success. It is assumed that female parental investment is increasing more rapidly with succeeding offspring than is male parental investment. Both sexes are being selected to maximize number of offspring as compared to parental investment and, as can be seen, this means that females are most fit when they produce O_f offspring, males when they produce O_m offspring. Since $O_m > O_f$, and since both sexes produce the same number of offspring, in this case males compete for females. Were $O_f > O_m$ then females would compete for males.

resource, but under certain special circumstances the roles can be reversed. For instance, in some species of fish it is the males which build nests, care for the young, and so on. In these cases, the males are the limiting resource sex, and, as the theory predicts, what we find is competition between the females for males, and associated adaptive mechanisms to aid the females in their competition.

As might be expected, there seems to be no one reason why the sex roles sometimes get reversed. In the case of the fish, with perhaps some plausibility it has been suggested that the reversal comes about through the mechanism of fertilization. Unlike the internal fertilization of land animals, where the female ends up pregnant, many fish have external fertilization, and, in particular, the male fertilizes *after* the female has laid her eggs. This means that the female can make her escape, leaving the male with the offspring. If it is in the male's reproductive interests that the offspring survive, and it obviously is, the male has no option but to raise the offspring himself. But, in turn, this means that the females battle over the males. (Dawkins, 1976, pp.167–9.)

So far, we have been considering selection as it would act on the sex making the lesser investment. But it will act also on the sex making the greater investment, for, remember, we suppose that selection acts at the level of the individual not the group. How might selection act to aid the individual making the greater investment, or how might it act to aid the individual which up to a certain point in time has made the greater investment? Clearly it might act to aid this individual to raise successfully the largest number of the best possible offspring. In particular, considering the individual's interactions with members of the opposite sex, we have at least two possible 'strategies' that might occur. On the one hand, the individual might be selected for characteristics which enable that individual to force members of the opposite sex into a greater investment in the offspring than would otherwise occur. On the other hand, we might have selection for characteristics which would enable the individual to complement its own genes with genes from the member of the other sex which would best enhance the future success of its own genes. In an entertaining popularization of sociobiology, Richard Dawkins (1976) colourfully refers to these two possibilities as the 'domestic-bliss' strategy and the 'he-man' strategy.[3] Let us take them in turn, assuming for ease of exposition that it is the female that is the limiting resource.

3.7. FEMALE REPRODUCTIVE STRATEGIES

In the domestic-bliss strategy, the female forces the male to make substantial investment before copulation. By the time copulation takes place the male is heavily 'committed' to the female, because he has had to build a nest, or feed the female extensively, or go through an elaborate courtship ritual, or the like. Basically, now it may not pay the male to desert, because the next female he will meet up with will also demand such prior effort. It may just be in his interest, rather than expending effort in trying to increase the absolute number of children he fathers, to divert his effort to attempting to see that those children he has already fathered come to maturity.

Of course, this all rather assumes that the next female to be met will demand prior effort. If she copulates immediately, then the male might just as well move on. We must therefore be able to show that such demand for prior effort by females could be held in a population. A group selection hypothesis could do this — it is obviously in the interests of the group as a whole if all the females gang up on the males — but this kind of hypothesis is forbidden. However, one can set up a simple model using a game-theoretic approach to show that the domestic-bliss strategy will work, even given just individual selection. Fortunately, since the details are fairly similar to those required in the restrained aggression model, there is no need for us here to go on to detailed exposition. (Dawkins, 1976, pp.163–5.)

It should, incidentally, be noted that as a kind of side-effect to the main selective forces at work here, one may also have some subsidiary forces centering on the possibility of cuckoldry. If a pattern of male parental care has evolved, then if the female does get deserted, selection may well have forced her to act at once to find a substitute father to help raise her offspring. But conversely, since any such step-father would be working to perpetuate the genes of strangers, one would also expect very strong selective pressures against such cuckoldry. This may be the reason why pregnant female mice abort when they smell strange males, and why male lions taking over a pride kill and eat the cubs. (Wilson, 1975a, p.85.)

It should also be mentioned that in this female strategy we may have a key as to why polygyny (single males with multiple females) is not uncommon in the animal world. In going with an already-mated male, a female is at least going with a 'proven' quality, particularly with respect to the provision of parental care, for instance defending a territory and providing food. It might be in a female's reproductive interests to forego exclusive access to an untried male for shared access to a tested male. (Wilson, 1975a, p.328.)

The other strategy open to the females is the he-man strategy. Here, in its purest form, the female abandons all hope of getting the male to help in the raising of offspring. What she wants from the male instead is the best possible genes for her offspring, or, less metaphorically, selection favours those females that are attracted to males who carry genes which best complement the females' genes. Conversely, selection favours males which carry such 'good' genes. Hence there is a strong selective premium on males being strong and agile and the like, and on a female being attracted to these qualities. It is in a female's reproductive interests that her children have these qualities too. Of course, although qualities like agility are of general value, there will also be the possibility of the development of characteristics which are not very helpful to the male. Suppose a characteristic was helpful to a male but perhaps has been developed beyond a point of usefulness. Nevertheless, it might still be selected because it is taken as a mark of attractiveness, and this in itself is of value to a mother, because then her sons will have the mark and thus have a better chance of reproducing.

The so-called 'he-man' strategy obviously gets very close to Darwin's sexual selection through female choice, what Julian Huxley (1938) later labelled 'epigamic selection', although whereas Darwin rather interpreted the peacock's feathers as being attractive to the peahen analogously to humans finding things beautiful, the modern sociobiologist perhaps rather interprets them as being attractive because they are indicative of the fact that the males are fit and that the female's male offspring will, in turn, be attractive. Thus, for all that the sociobiologists invariably use anthropomorphic language in describing and labelling these phenomena, essentially perhaps they are less open to criticism than Darwin on the score of anthropomorphism. The so-called 'domestic-bliss' strategy perhaps was not really covered by Darwin's sexual selection, although no doubt usually one would also get sexual selection through male combat as males vied to show females that they would be better providers and parents than other males. And obviously also the he-man strategy could well be combined with some male combat: females being attracted to the winners, because then they would have more chance of male children who would be winners.

Note that in the operation of these various female strategies, there will be constant tension as males try to appear fitter than they really are and females try to discriminate between those that are really fit and those that just seem to be. About this, in a passage which we shall later see has become somewhat notorious in the sociobiological controversy, Wilson has written as follows:

Pure epigamic display can be envisioned as a contest between salesmanship and sales resistance. The sex that courts, ordinarily the male, plans to invest less reproductive effort in the offspring. What it offers to the female is chiefly evidence that it is fully normal and physiologically fit. But this warranty consists of only a brief performance, so that strong selective pressures exist for less fit individuals to present a false image. The courted sex, usually the female, will therefore find it strongly advantageous to distinguish the really fit from the pretended fit. Consequently, there will be a strong tendency for the courted sex to develop coyness. That is, its responses will be hesitant and cautious in a way that evokes still more displays and makes correct discrimination easier. (Wilson, 1975a, p.320.)

In concluding this discussion of sociobiological work on sexual selection, it is perhaps not out of place to end on the same note of caution as was struck at the end of our discussion of sociobiological work on aggression. The facts of the animal world do seem to point unequivocally to the operation of some kind of sexual selection, broadly conceived. The differences between males and females are just too common and too striking to be otherwise. Moreover, as I have tried to intimate through the discussion, many of the particular suggestions of the sociobiologists seem to find empirical confirmation. For instance, Wilson shows that, generally speaking, as one might expect, right through the animal world males seem conspicuously absent when any hard work with the children is required; and, moreover, males tend to be sexually promiscuous. On the other hand, where really intensive care is required, males can be 'persuaded' to take an active role. For example, in the case of many species of birds the males help with child rearing.

Nevertheless, for all of these positive facts these are early days, and in many respects we hardly have hard-tested theory. For instance, models can be set up to show that individual selection can promote and maintain female behaviour that will cause males to provide extensive parental care; but, as in the case of aggression, the models are as yet crude and the appropriate figures are chosen to get the right results, not because they are proven fact. But perhaps one should not ask for more at this stage.

3.8. PARENTHOOD

We come next to the results of sex, namely offspring and the related responsibility, parenthood. Here, surely, one might think, selective interests coincide. Two parents might not have the same interests, and therefore can be expected to try to maximize their own reproductive interests at the expense of the other; but surely the interests of parents and children coincide. However, whilst there is obvious truth in this initial impression, matters are not quite straightforward, as Trivers has been ready to point out. Let us see why this is so. (Trivers, 1974.)

Individual selection supposes that one will strive to maximize the opportunities for one's own genes to reproduce – if one does not, then one's genes will be pushed out by the genes of those organisms that do. Now, considering sexual organisms, if one is the only child of two parents, (and will stay that way) then the interests of the parents and the offspring coincide. The only way that the parents can pass on their genes is through the child. However, of course, being an only child is a rare phenomenon (in the animal world). Usually a parent (for convenience let us just stick with the mother) will have more than one offspring. This means that, genetically speaking, the parent and any particular offspring have different interests. The parent wants to maximize the number of her genes, the offspring wants to maximize the number of its genes. More precisely, trying to get away from false implications of conscious intention, the genes of the mother are promoting characteristics that will lead to their own replication; the genes of the offspring are promoting characteristics that will lead to their own replication. But, whereas the parent's genes are distributed 50% in each offspring, the offspring's genes are concentrated 100% in itself (this latter claim must be qualified in a moment). Hence, the two sets of genes can and will promote characteristics that lead to conflicting behaviour.

Invoking once again the notion of parental investment, we can say that the mother has a fixed total parental investment, which she will try to apportion out amongst her offspring to maximize the number that survive to reproductive maturity. This means that almost inevitably she will limit the amount that she will give to any one child. On the other hand, the child wants to maximize its own chances, that is its genes will promote characteristics to maximize their chances of replication, and so the child in turn will try to exceed the limited amount of investment that the parent offers to it. And all of this adds up to conflict – at least it does after the mother has given the offspring the share of care which it is in the mother's reproductive interest to give.

Actually matters are a little more complex than suggested so far. To this point, it might seem as if the offspring is going to hound the parent all of its life for help. However, there is an important reason why this would not be so, namely that any offspring is also related to the rivals for its parent's investment: that is, the child is related not only to itself but to its siblings also. Drawing on theory developed in a pair of seminal papers by W. D. Hamilton (1964a,b), Trivers points out that any individual is 50% related to its (full) siblings. This means that it can be in an organism's reproductive interests to see a sibling survive, simply because the organism shares genes with the sibling and therefore inasmuch as the sibling survives and reproduces it passes on the organism's own genes (more strictly, it passes on *copies* of the organism's own genes, which is all that the organism itself passes on anyway).

Of course, an organism is more closely related to itself than to its siblings (100% to 50%). This means that, generally speaking, we can expect a three-stage process in child care. An initial period when it is in the interests of both parent and child that the parent invest in the child. A final period when it is in the interests of both parent and child that the parent invest in other siblings: the loss to the child is more than compensated by the gains of 50%-related siblings (it is easy to see that it must do the siblings more than twice

as much good to get a unit of care as it does the individual child). And then
we have a period in the middle. It is in the parent's reproductive interest to
invest in other children, but the child itself is still getting enough out of the
parent's interest for it not yet to be of more worth to the child that the
interest be directed to others. In other words, this is the period between when
a unit of help does more good for other siblings than to an individual but not
twice as much good, and when the unit does do at least twice as much good
for other siblings. In this period, we expect parent–child conflict. (Clearly
the kind of model being presupposed here is of a growing child, which gets
less and less from each unit of care, and of younger siblings who would benefit more from such units.) (See Figure 3.3.)

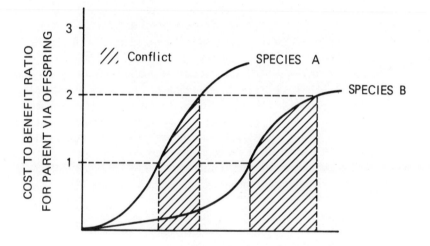

AGE OF OFFSPRING

Fig. 3.3. (From Trivers 1974, via Barash 1977.) Here we see graphically how parent-
offspring conflict arises when the parent cuts off aid. In both species the conflict arises
when the parent's reproductive interest lies in helping other children, but this is not yet
the interest of the offspring. Species A terminates care before Species B.

Now with this theory we are getting close to the notions of kin selection
and of inclusive fitness — notions which lie at the heart of some of the most
impressive work done to date by sociobiologists. This work, both the theory
and the evidence, will be discussed in the next section. Remembering, therefore, that we are here not discussing indirect evidence for the claims on which

Trivers' analysis of parent–child conflict rests, let us note that as before although there may not be much quantified evidence for Trivers' claims, what evidence we have makes them seem broadly plausible. At least, the evidence does not rule them out, as it rather does for rival analyses, particularly those based on group selection.

Indeed, if we look at matters from a group selective viewpoint, one would expect there to be little or no parent–offspring conflict, at any time. Both parent and offspring are members of the same species, and so selection should act to harmonize the desires of parents and children. And, yet, it is one of the most widely documented facts of the animal world that parents and children differ at times in their wishes, and that these differences become particularly acute at the times at which parents are breaking off their care. As Trivers writes: "Weaning conflict in baboons, for example, may last for weeks or months, involving daily competitive interactions and loud cries from the infant in a species otherwise strongly selected for silence." (Trivers, 1971, p.251.) The only alternative explanation, given individual selection, seems to be that here we have a general and widespread inefficiency of nature: obviously most of the time parents and children agree, and it is just at the point of changeover that things get out of focus, but things are not sufficiently deleterious to the individuals involved that selection would do something about it. However, this is always an explanation of the last resort, and the simple fact of the matter is that weaning does cause disruptions of a kind and magnitude that would, under normal circumstances, be attended to by selection. Hence, the plausibility of the sociobiological explanation.

3.9. ALTRUISM

We come to the final topic, or cluster of topics, to be discussed in this chapter. Altruism, roughly speaking, means doing something for someone else. More precisely, from a biological viewpoint, it means doing something to help the reproductive chances of someone else, even though apparently this entails the diminution of one's own chances. It is obviously the *sine quo non* of social behaviour: indeed, it might be felt that in some sense it is involved in the definition of what we mean by 'social' behaviour. But, without wanting to get bogged down in semantics, we can certainly say that altruism as characterized above has been of intense interest to sociobiologists. Indeed, Wilson goes so far as to claim that it constitutes "the central theoretical problem of sociobiology". (Wilson, 1975a, p.3.) And, of course, the reason why he would want to say something like this is not hard to see. If one is an individual selectionist, and the sociobiologists to a person are, then altruism presents one with a major paradox. Unlike group selection, individual selection claims that it is absolutely impossible that selection promote characteristics which are not of benefit to the individual. But altruistic characteristics seem to be precisely those that are not of benefit to the individual. In short, individual selection seems to bar altruism.

We cannot deny the facts of altruism. Evolutionists from Darwin on have recognized that it is a widespread phenomenon in the animal world, perhaps reaching its climax in the sterile castes found in insects, where workers devote themselves entirely to others. How then is the sociobiologist to escape from the dilemma? Fairly obviously, by showing that for all appearances to the contrary altruism benefits the reproductive interests of the individual causing the altruism. As one sociobiologist has written in a discussion of altruism: "As pointed out by various authors . . . there is ultimately no such thing as biological altruism. Obviously, 'altruism' as defined here is ultimately selfish in leading to the spread of alleles like those of the performer in the population." (West Eberhard, 1975, p.7.) Leaving, until later, discussion of how profitable it is to speak of 'selfishness' at this point, and noting also that altruism might not necessarily be of direct benefit to the performer, the main idea is clear: altruism must be interpreted in terms of the reproductive interests of the individual causing the altruism (here assuming that an individual's genes can cause its own altruism).

Viewing matters in this light, we find that sociobiologists have suggested three major possible causes of animal altruism: kin selection, parental manipulation, and reciprocal altruism. Let us take these in turn.

3.10. KIN SELECTION

Kin selection starts from the fact that we are related to others. (Hamilton, 1964a,b.) This means that we share some genes with others, and that, hence, inasmuch as our genes have been selected precisely because of their ability to cause characteristics which will ensure the genes' replication, it is in our reproductive interests to see that those who share our genes reproduce, because they are then making copies of the genes that we have. Another way of putting matters is to say that only those genes that reproduce, persist; and it does not matter whether this reproduction is done directly, or by proxy. But, whichever way things are put, what is implied is that it might just be in our own reproductive interest to have relatives reproduce. In short, it might be worth my while to be altruistic towards my relatives, because they will then pass on copies of my genes. (We can therefore coin the notion of 'inclusive fitness', which is the fitness of an individual in its own right, together with its influence on fitness in non-descendent relatives.)

We have already met a version of this argument in the context of parent—offspring conflict, where it was seen that there comes a point when it pays an individual not to take the food out of the mouths of its siblings. However, it can readily be seen that this is only a particular instance of a more general theory. Any help (*i.e.* reproductive help) towards any relative is of reproductive value to me. Of course, this last statement must be qualified — it is more in my interest to help, say, my full brother than a first cousin, for as can be seen from Mendel's first law the former shares 50% of my genes whereas the latter shares only 12.5% of my genes. Indeed, as can readily be seen, one can set up a simple formula. If it pays my brother more than double what I would gain from certain effort (or what I lose if the effort is expended without return), then it pays me to help him. Again, if it pays my cousin more than eight times the effort's value to me, then it pays me to help. And, more generally, altruism towards relatives is worthwhile if the ratio of gain to loss in fitness (k) exceeds the reciprocal of the average co-efficient of relationship of benefiting relatives (r).

But other than the rather special case of older siblings standing aside for the benefit of younger siblings, does kin selection promoting altruism ever get a chance to take effect in the wild? It is suspected that it may be fairly prevalent, particularly when near relatives tend to live close together. A nice example may be provided by wild turkeys. (Wilson, 1975a, pp.125—6.) Male wild turkeys form ever-inclusive groups (*i.e.* nested sets): one has bands of brothers, which come together in groups, and which in turn come together in yet larger flocks. Status in all of these relative groups is very clearly defined and is established by wrestling. Brothers wrestle together to establish top turkey, then the bands of brothers wrestle, and so on. Success in wrestling pays very high reproductive dividends, for it is only the cream of the top bands that get to mate with the hens — all others do without. Kin selection is a great help in explaining this phenomenon: why, for example, a turkey may be prepared to wrestle for his band, even though success will only open up breeding opportunities for his higher ranking brothers. Having been beaten by his brother, his own chances of reproducing are much diminished. However, making the best of a bad job, if his brother reproduces, at least some of his genes will be passed on. Hence, altruistic fighting for the brother pays. (More precisely, since any effort on his own behalf is almost bound to be wasted, any effort expounded helping the brothers' chances is bound to outweigh the mere 50% relationship – presumably, as one fans out in the larger groups one still has relationships, albeit diminishing. Of course, no conscious intention is implied.)

It was mentioned above that the area of the animal kingdom where altruism is shown in its purest and most absolute form is the social insects, particularly the Hymenoptera (ants, bees, and wasps), where whole castes of sterile females devote their time exclusively to the well-being of their mother (the Queen) and their siblings. It might therefore be expected that the sociobiologists would turn their attention to this area, and indeed we find that the most exciting, not to say daring, application of the notion of kin selection occurs at this point.

It has been suggested by Hamilton (1964a,b) that the key to the phenomenon of sterile worker altruism, made even more striking by the fact that it is believed to have evolved several times independently in the Hymenoptera and only once elsewhere (the Termites), lies in the peculiar way in which sex is determined amongst the Hymenoptera. In particular, females are diploid having both father and mother, whereas males are haploid having only a mother. If the fertilized Queen in turn fertilizes an egg she has a daughter; otherwise she has a son. This means that sibling daughters of a Queen, fertilized by a single male, are more closely related to each other than they would be to any of their own daughters! Consider: If the females have the same father, then that gives them a 50% relatedness, for the father being haploid will give each diploid daughter exactly the same genes, and one can add another 25% relatedness because the females share a diploid mother who gives them the other half of their genes, and by Mendel's first law any two daughters will have half of these maternally donated genes in common. In other words, sibling daughters will be 75% related (i.e. share 75% of their genes).

On the other hand, mothers and daughters will share only 50% of their genes (the other 50% of the daughters' genes coming from the father). Hence, argues Hamilton and those sociobiologists who agree with him, because being a sterile worker and not a reproductive is a function of the environment and not the genes (e.g. is caused by not being fed some special nutrients), we can see how it would 'pay' a female to forego entirely her own reproduction and devote her time entirely to raising fertile sisters (of course, this might require raising other worker sisters): fertile sisters will spread more of the worker's genes than fertile daughters. More precisely, genes in the worker which bring about altruistic behaviour will be at a selective advantage to genes which do not.

Apart from the sheer elegance of this explanation and apart from the above-mentioned near confinement of worker sterility to the Hymenoptera,[4] there are other obvious facts in its favour. For instance, one does not find male workers, and Hamilton's explanations show why. A male is no more closely related to siblings than it is to daughters (it has no sons). Of course, this is all a bit informal, hardly excluding rival explanations, and so, recently, through an ingenious argument, Robert Trivers and Hope Hare (1976) have tried to put Hamilton's hypothesis to a fairly rigorous test. They point out, following Sir Ronald Fisher, that normally sexuality implies a 50:50 distribution of males to females. The argument behind this is analogous to the argument for gene balance based on rareness, given in the last chapter. Any deviation from a 50:50 ratio would give a selective advantage to an organism producing more of the rare sex; and thus the ratio would be brought back into balance. (Strictly speaking, what one has is not a 50:50 numerical ratio, but a 50:50 ratio of the *effort* to produce males and females. If a male requires less effort to produce than a female, then selection favours producing more males.)

However, argue Trivers and Hare, in the case of Hymenoptera the normal sex-ratio should not obtain, or at least not necessarily obtain. If the Queen controls the nest, then a normal 50:50 ratio should hold. However, if as usually happens, the workers control the nest, then there should be a bias in favour of females. Because a female is more closely related to her sisters than to her brothers, she can maximize the spread of her own genes by raising more fertile sisters than she raises brothers. In particular, one can show that the ideal ratio of males to fertile females (from the workers' viewpoint) is 25:75 (more strictly, the ratio is the effort to produce organisms, which Trivers and Hare correlate to body weight). And, bringing their argument to a seemingly triumphant conclusion, by appealing to a wide range of empirical evidence, Trivers and Hare argue that this is precisely the kind of ratio which does in fact obtain. Moreover, in certain special cases where the Queen can control her nest, specifically when she relies on unrelated 'slave' workers, they suggest one finds more normal sex ratios: as expected.

Unfortunately, although for a while this argument of Trivers and Hare was taken as triumphant vindication of kin selection in particular and sociobiology in general, recently its worth has been seriously questioned. For instance, it is crucial to the argument that Queens have only one mate. If they have more than one, then at once the worker-sisters' close relatedness vanishes. But, it has been suggested, in reality Hymenoptera Queens do often mate repeatedly. (Alexander and Sherman, 1977.) Thus Trivers and Hare's argument fails; although the critics do not deny that the Hymenoptera produce a suspiciously large number of fertile females. Consequently, in the place of a kin-selection argument they suggest another of Hamilton's mechanisms, which will produce fewer males. This mechanism, 'Local Mate Competition', (Hamilton, 1967), starts from the fact that if the relatives of a certain sex compete for mates, then it is in their parents' reproductive interest to produce fewer of them. For instance, if two brothers compete for every mate, and one brother could alone fertilize the mates, then the parents have wasted effort in producing two brothers. In the case of the Hymenoptera, argue the critics, we do get such competition between males, and thus we get the biased sex ratios. But not for the reasons that Trivers and Hare suppose.

We are talking about a fast-moving field, and undoubtedly the last word has not been said about this and related matters. It could well be that neither explanation of the sex ratios is exclusively right or exclusively wrong. However, in drawing this section to a close, the reader will by now realize that the notion of kin selection is a very powerful one; that it has exciting potentialities; but that at present its full worth is far from clear. (Wilson, 1975a, was one who initially was very enthused by Trivers and Hare's results. Since Alexander and Sherman's critique, he has tempered his enthusiasm somewhat; but against Alexander and Sherman he suggests that in many ant species of the kind studied by Trivers and Hare, local mate competition is not likely. See Oster and Wilson, 1978.)

3.11. PARENTAL MANIPULATION

It is suspected by some sociobiologists (especially Alexander, 1974) that a form of altruism can be evolved through selection, not so much because it is in an individual's reproductive interests to help its close relatives, but because it is in the interests of the individual's parents that it offer help, and the parents have been able to manipulate the individual into offering that help. If we cast our minds back to the parent—offspring analysis, it is easy to see how such a state of affairs might possibly come about. Suppose a parent has a number of children, say five. Suppose that it is impossible for the parent to raise all five: indeed, even if one child is abandoned, no more than three of the other four will be raised. Suppose, however, that the parent is in a position to manipulate one of the children, so that this child becomes an altruist towards its siblings, and thus all four of the others survive. Clearly, in such a case, any genes causing such manipulation on the part of the parent will be strongly favoured by selection. (Although see Dawkins, 1976, pp.146—7.)

It may seem at first that there is not much difference between parental manipulation and kin selection; but there is, in fact, an important difference. In kin selection, one individual helps another individual because they are related: the second individual is passing on the first individual's genes. In parental manipulation, one individual is forced to help a second individual for the sake of a third individual. That the first and second individuals share genes is quite incidental. Of course, as a matter of fact the first and second individuals *will* share genes, so possibly when one has parental manipulation kin selection comes into play also. And, certainly, it is not easy to distinguish between cases of parental manipulation and kin selection. If one has siblings helping, as happens in certain species of bird, it may be a function of parental manipulation – but then again it may be a function of kin selection.

It has been suggested, particularly by the critics of Trivers and Hare (!), that it is parental manipulation which is the main causal factor at work in the Hymenoptera: after all, when setting up the nest the Queen does make workers rather than reproductives, by virtue of what she feeds her first offspring. (Alexander, 1974.) However, as we have seen, this is all very much a matter of unresolved controversy at the moment. More promising for the hypothesis of parental manipulation, perhaps, are the fairly frequent cases in the animal kingdom where one offspring is fed to the others. The phenomenon of so-called 'trophic' eggs, which are used as food by siblings, is common in the insect world. And even in the higher vertebrates one gets cases of cannibalism within the litter. Of course, there may come a point in the life of an organism when it would pay it to lay down its life for its siblings; but at the very early stages of development, when an organism is about to be fed to its siblings, it would hardly seem that the probabilities are so clear-cut as to make such an act of self-sacrifice worthwhile.

Incidentally, if the parental manipulation claim has any truth, then one might get an interesting variant on the parent–offspring conflict scenario. Normally, conflict comes about because offspring want help that the parents are loathe to give. But if the parents use excess help as a way to bind offspring to them, then it might well be in an offspring's self-interest to rebel. In other words, the conflict could go in a reverse direction, with the parent offering, and the child rejecting, help. Of course, it is not being supposed that any of this Machiavellian behaviour is consciously planned. The behaviour is supposed to be a function of the genes, and indeed there might be a selective value in the body's functioning in ignorance. (Trivers, 1974.)

3.12. RECIPROCAL ALTRUISM

Thirdly, and finally, amongst the suggested possible causes of animal altruism, we have reciprocal altruism. (Trivers, 1971.) This, if it has any truth in it, is a much more wide ranging mechanism than the previous mechanisms for altruism, in that it can occur between perfect strangers and even between members of different species. Again we must turn to the thought of Trivers to find details. Suppose we have two individuals, A and B, and that each stands in danger of drowning (Trivers uses human beings in his example, but that is not necessary). Suppose that unaided, an individual has a 50% chance of drowning, but that if aided the chance of drowning drops to about 5%. Suppose in addition that would-be rescuer and rescued always drown or survive together, and that everyone at some point faces the risk of drowning (i.e. needs help). Clearly, if everyone in the group is also at some point prepared to lend a helping hand, the drowning risks of a member of the group drop from 50% to about 10%. In the long run, everybody gains.

To a group selectionist, this would be all that there is to the matter; but for the individual selectionist, important questions remain. Why should one put one's life at risk for the sake of another? Obviously, because one hopes in turn for help when one is oneself in trouble (that is, those genes which put themselves at risk, reap selective rewards — as always, no conscious intentions are supposed). But then the question arises: Why not cheat? Why not accept rescue for oneself but ignore the pleas of others? Here, one has to suppose some faculty of remembering, but then the answer is simply that individuals remember cheaters and refuse to help them. In other words, one helps those who have shown a willingness to help others, or who at least have not shown an unwillingness to help others. Thus one gets, on a kind of principle of enlightened self-interest, the spread and maintenance of altruism through a group, even a group of non-relatives; although, probably one will always have a certain amount of cheating, particularly in a big, fluid population, because it will take a while for one's cheating nature to become generally known.

The difference between the reciprocal-altruism and kin-selective case is that in the former one expects some direct return for oneself whereas in the latter one does not expect any such direct return for one's own altruism: the return is in seeing reproductive success of genes shared by oneself. As the reader must have concluded, there are very direct links between this approach taken by Trivers in his analysis of reciprocal altruism and the approach taken by Maynard Smith in his analysis of animal aggression. And indeed, although Trivers does not invoke all of the formal apparatus of game theory, he does explicitly liken the situation of the reciprocal altruist to that faced by a player in the Prisoner's Dilemma, a favourite subject of game theorists.

But does one find in nature any actual instances of animal altruistic behaviour which seem suitable subjects for a reciprocal altruism analysis? Trivers discusses in detail one case which *prima facie* seems most convincing: cleaning symbioses in fish. The members of certain species of fish, shrimps and the like, clean parasites off the members of other species of fish. On the one hand, the cleaners gain by getting a good meal; on the other hand, the cleaned gain by losing parasites which, if left untended, can lead to all sorts of deleterious sores and diseases. But what is truly remarkable about such situations is that, although the cleaners would make nice nutritious meals for the cleaned fish, only rarely does it happen that a cleaner ends its task by becoming someone's supper. Indeed the cleaned fish will usually go to considerable pains to ensure the safety of the cleaners, even when they are themselves threatened. (Feder, 1966; Maynard, 1968.)

Evidence shows without a doubt that the behaviour of cleaners and cleaned is under the control of the genes. Indeed, there are some clearly documented examples of some really vicious fish having been reared in isolation, but immediately showing appropriate protective care when faced with cleaners. Moreover, the selective advantage of the various kinds of behaviour exhibited has also been demonstrated. The large fish tend to go to various stations to be cleaned, and if the cleaners are removed artificially (*i.e.* if it is made as if the cleaners were being eaten), the would-be cleaned fish very rapidly develop sores and other disabilities. And, finally — the almost inevitable mark that something is controlled by genes and is of selective advantage (because otherwise it would end pretty quickly) — it is found that other species of fish take advantage of the situation. Some fish pretend to be cleaners, but in fact merely take advantage of the situation, rushing in to bite chunks out of the unsuspecting larger fish waiting for their grooming!

In this particular example, it is clear that kin selection cannot be at work. There is no genetic relationship between the cleaners and the cleaned. Therefore, it is argued by Trivers that one has to invoke some kind of reciprocal altruist explanation. Certainly, nothing very much is being quantified; but it is felt that only a kind of enlightened self-interest explanation will suffice. Conversely, this implies that animal altruism can be promoted by individual selection, even across species' barriers.

With this discussion of reciprocal altruism, we draw to the end of our excursion through the ideas and results of the sociobiologists as they apply to the animal world. Avowedly, I have not tried to cover everything, but the reader should now have sufficient of a flavour of the enterprise. Let us therefore turn at once to the human world.

NOTES TO CHAPTER 3

[1] For ease of exposition at this point, I use a simplified model presented by Dawkins in his popularization of sociobiology (Dawkins, 1976).

[2] For the moment, let me speak informally. I do not want to imply conscious intention by my use of the term 'want': merely that something is in an individual's reproductive self-interests.

[3] Frankly, these names strike me as being as much silly as colourful. But one of the complaints about sociobiology has been its language and use of metaphor, and I am anxious not to dodge this issue.

[4] Oster and Wilson (1978) suggest that the reason for the sociality of the termites might lie in their necessarily living close together because they need to pass on to each other certain symbiotic intestinal flagellates through anal feeding.

CHAPTER 4

HUMAN SOCIOBIOLOGY

We come now to the area of major controversy: the application of the ideas sketched in the last chapter to our own species, *Homo sapiens*. In fairness to many of the sociobiologists (and to sociobiology in general) it is important to begin with a qualification: One can do animal sociobiology without taking any real interest in human sociobiology, or indeed whilst denying that socio-biology can be extended in any legitimate way to humans. Hence, whatever one's convictions, one ought not rush in praise or condemn sociobiology as such on the strength of its relevance to humans.

In fact, what one seems to find from sociobiologists is something of a spectrum of reactions when it comes to humans, with, generally speaking, American thinkers being much more ready than British thinkers to see hu-mans as sociobiological organisms. Thus, for example, at one extreme, Trivers writes: "The chimpanzee and the human share about 99.5 per cent of their evolutionary history, yet most human thinkers regard the chimp as a mal-formed, irrelevant oddity while seeing themselves as stepping-stones to the Almighty. To an evolutionist this cannot be so. There exists no objective basis on which to elevate one species above another." And he continues: "Natural selection [the non-random differential reproduction of genes] has built us, and it is natural selection we must understand if we are to comprehend our own identities." (Trivers, 1976, p.v.)

Then, next in line, we have someone like Alexander: "A Darwinian model seems to me to be established beyond doubt as an appropriate hypothesis for the background of a significant, if not the major, fraction of the proximate behavioral tendencies and motivations of humans." (Alexander, 1977b.) Also, there is someone like Wilson. Clearly he thinks that sociobiology is in some broad sense applicable to humans, but just how far applicable is not easy to make out. At one point Wilson admits that: "It is part of the conven-tional wisdom that virtually all cultural variation is phenotypic rather than genetic in origin." (Wilson, 1975a, p.550.) And he himself immediately con-cedes that "the genes have given away most of their sovereignty". But then it becomes clear, both by admission and approach, that Wilson still sees the human genes as having a substantial causal role in human behaviour. More-over, Wilson rather complicates matters by introducing something which he calls the 'Multiplier Effect'. Wilson writes: "Social organization is the class of phenotypes furthest removed from the genes. It is derived jointly from the behaviour of individuals and the demographic properties of the population, both of which are themselves highly synthetic properties. A small evolutionary change in the behaviour pattern of individuals can be amplified into a major social effect by the expanding upward distribution of the effect into multiple facets of social life." (Wilson, 1975a, p.11.) In other words, Wilson believes that small genetic effects can have major behavioural consequences, and it is clear that he thinks this is particularly so in the case of humans.

Crossing the Atlantic, we find that Dawkins (1976) is much less ready to apply sociobiology to humans. Indeed, he believes that in most important respects humans have escaped their biology, and he proposes for humans his own theory of cultural evolution, where 'memes' which are kinds of intellectual units have replaced genes. Nevertheless, in discussing the theories and facts of animal sociobiology, Dawkins is not above the occasional reference to the possible human implications of what he is writing. And, finally, right at the other extreme we have Maynard Smith (1972) who flatly dissociates himself from human sociobiology and who denies that his ideas have any relevance to the human case.

So much for general positions: Let us turn now to some of the details. In this chapter I shall follow the order of topics as presented in the last chapter, but of course dealing exclusively with humans. My aim here is to give exposition. Critical comment will be reserved for later.

4.1. AGGRESSION

Right from the beginning, the topic of human aggression has attracted the interest of evolutionary biologists, both as cause and as effect. As is well known, although in the early years many people rejected Darwinism for extra-scientific political reasons, others were attracted to Darwinism for extra-scientific political reasons. They thought that his biological ideas could be used to support all kinds of human aggressive practices — the high point coming with John D. Rockefeller's 'Darwinian' justification of his cut-throat, bullying business ventures. (Himmelfarb, 1968.) Today, perhaps because so many of these extrapolations have fallen into disrepute, there is somewhat of a tendency to deny that Darwin had any real hand in fathering beliefs that humans are by nature aggressive and that this aggression may have had a significant role in their history. The origins of 'Social Darwinism' are usually laid at the feet of Herbert Spencer. Actually, however, even though Darwin did certainly at times disavow biological justifications for 'Might is right' doctrines, he was fully committed to the biological centrality of human aggression. He saw such aggression as having been of key evolutionary importance to humans, although admittedly as always with biological phenomena he was prepared to explain it causally in terms of the inheritance of acquired characteristics as well as the natural selection of the more aggressive (which aggression, or rather successful aggression, he believed linked to brain size). Moreover, Darwin was not beyond drawing consequences for the present: "The more civilized so-called Caucasian races have beaten the Turkish hollow in the struggle for existence." (Darwin, 1887, 1, p.316.)

Today, we tend to cringe from such crudely chauvinistic inferences, but the biological bases of human aggression have continued to attract attention. In fact, as pointed out in the last chapter, a rather curiously self-depreciating general impression has arisen. On the one hand, following the likes of Lorenz, with species other than *Homo sapiens* all is seen — if not as sweetness and light — as relatively harmonious and certainly non-lethal. "When, in the course of its evolution, a species of animals develops a weapon which may destroy a fellow-member at one blow, then, in order to survive, it must develop, along with the weapon, a social inhibition to prevent a usage which could endanger the existence of the species . . ." (Lorenz, 1966, p.113.) On the other hand, following people like Raymond Dart, humans are seen as unrestrained and uninhibited killers, not only of other species but also of their own. Biologically, we are all descended from Cain. Thus, Lorenz again: "There is only one being in possession of weapons which do not grow on his body and of whose working plan, therefore, the instincts of his species know nothing and in the usage of which he has no correspondingly adequate inhibition. That being is man." (Lorenz, 1966; also Dart, 1953.)

We learnt in the last chapter that, with respect to animals, the sociobiologists find both the facts and the theories of their predecessors highly suspect. They deny that animals are so very restrained towards their fellows and they deny group selective hypotheses like that of Lorenz. Similarly, in the case of humans they deny facts and theories. For a start, they suggest that humans are nothing like as aggressive as they are often portrayed, certainly relative to other species. Wilson writes that "murder is far more common and hence 'normal' in many vertebrate species than in man." (Wilson, 1975a, p.247.) Indeed, Wilson suggests that compared to lions and the like, we humans are a very friendly lot indeed. And also, as might be expected, Wilson and other sociobiologists are loathe to explain human behaviour in terms of group selection. But none of this is to deny that sociobiologists do see aggression in humans, that they think it of major importance, or that they attempt to use their animal sociobiological findings to illuminate it. Socio-biologists recognize and are concerned with human aggression; they think that it is a significant facet of human behaviour now and in the evolutionary past; and moreover, with some significant exceptions to be noted, they think that their work in animal sociobiology is pertinent here.

For a start, we have seen that Wilson believes that much animal aggression is adaptive, being directed towards the gaining of limited resources. "Non-sexual aggression practiced within species serves primarily as a form of com-petition for environmental resources, including especially food and shelter." (Wilson, 1975a, p.243.) And he sees it as being triggered by a number of factors, most particularly by the arrival of strangers. "This xenophobic prin-ciple has been documented in virtually every group of animals displaying higher forms of social organization." (Wilson, 1975a, p.249.) In an identical manner, Wilson sees aggression in humans, not as some murky trait showing our essentially blood-lust character, but as something widespread and of great adaptive significance for the survival and reproduction of the individual, most specifically when such an individual is faced with competition for limited resources, such as food or just general living space. Moreover, although Wilson expresses indifference as to whether aggression is genetic or learned, it is clear that in the most essential sense Wilson thinks aggression genetic or innate. Certainly, he likens our responses under stress to those of other organisms where any aggressive responses most definitely are genetic. Thus cats and Norway rats get really aggressive and bizarre under extreme crowding condi-tions. And there are "some clear similarities, for example, between the social life of [restricted] rats and that of people in concentration and prisoner-of-war camps, dramatized so remorselessly, for example, in the novels *Anderson-ville* and *King Rat*". (Wilson, 1975a, p.255.) Furthermore, human response to strangers parallels closely the responses of other animals: intuitively, we put up barriers to strangers, foreigners, outsiders, and the like. They are the Yids and Wops and Spicks.

So much for aggression in the present. What about the past? Here, it is clear that Wilson is in general sympathy with Darwin, seeing aggression as having played a key part in the human evolutionary past. As also is the socio-biologist Richard D. Alexander, who argues that somewhat paradoxically much of the human ability to live socially together (which Alexander sees as being underlyingly genetic) is ultimately an outcome of aggression. It is often suggested that humans started to band together and to evolve rapidly as a function of hunting — ancestral humans needed to work together intelligently in order to secure prey far bigger, fleeter, and stronger than they. However,

Alexander finds this explanation alone unconvincing, particularly when coupled with the deleterious effects of group living, such as the greater likelihoods of disease and parasites. He believes that sociability and intelligence, not to mention aggression, are adaptive responses to predators competing for similar resources, namely other bands of humans. "I suggest that, at an early stage, predators became chiefly responsible for forcing men to live in groups, and that those predators were not other species but larger, stronger groups of men." (Alexander, 1971, p.116.) It should be added that Alexander does not just suppose a straight survival-of-the-fittest cause for intelligence growth, but that he invokes kin selection also. In particular, Alexander suggests that there would be selection for ability to discriminate between kin (friends) and non-kin (enemies): this ability being intelligence.

So far, we have just been considering aggression without restriction. What about human aggression with restrictions? Again, it is to Alexander that we might most profitably refer. Contra Lorenz and others, "man, who clearly has the most elaborate and complicated selfish inhibitions to aggression in the animal kingdom, may also possess the ability to preserve his species from the destructiveness of his aggression". (Alexander, 1971, p.114.) And, Alexander makes very clear that in his opinion this ability to help the species is only a function of the individual human ability to help itself. In particular, Alexander pursues the same kind of reasoning as do those who apply game theory to animal aggression: all-out attack if one is certain of winning is fine; but if the probabilities of success are much lower, or if the probabilities of cost whatever the outcome are much higher, then a strategy of restrained aggression may be the best policy. "The worst kind of animal on which to press an attack with a low probability of gain, after all, is one that has lethal weapons; if you kill him but receive a mortal or disabling wound in the process, you are certain to lose rather than gain. Lions and tigers . . . seem not different from nations with nuclear weapons in this regard." (Alexander, 1971, p.114.)

Furthermore, in support of his position on restrained human aggression, Alexander refers to anthropological data on various 'primitive' tribes: often these go through all kinds of threatening and bluffing procedures with opponents. Rather than seeing ritualized play, in itself somehow inhibiting violence, Alexander suggests that we may well be watching ferocious opponents testing out the opposition, before deciding that all-out conflict is not in their self-interest. (Since Alexander tends to see small bands as kin, there is no suggestion of a group-selective mechanism at work here.)

Obviously, although he does not set matters up in formal terms, Alexander's analysis here runs parallel to the analysis of Maynard Smith for animal aggression – Maynard Smith having taken matters one step further and shown exactly how such a mechanism might work. However, we have a rather curious gap at this point. Alexander, on his part, does not refer to Maynard Smith's work: admittedly, he could hardly have done this at first, because his earliest work pre-dates Maynard Smith's. But even later he does not seize on the game-theoretic approach as supportive. And, as intimated at the beginning of this chapter, Maynard Smith wants no part of applications of his approach to humans. Despite the fact that he admits that game theory was modelled originally on human behaviour, despite the fact that he admits that some primitive peoples go through aggressive posturing in a manner similar to animals, despite the fact that he wants to apply his theorizing to primates, despite the fact that he even refers to work where apparently such an approach as his holds for neurotic humans, Maynard Smith categorically denies that he has pointed towards a legitimate, game-theoretic, sociobiological approach to human behaviour. "There may or may not be physiological similarities between human and animal aggression, but nothing I have said in this essay [applying game theory to animal aggression] is intended as evidence for such similarity." (Maynard Smith, 1972, p.25.) Any similarity is of a different kind. "I think there is often a logical similarity between the role of human reason in optimizing the outcome of a conflict between men, and the role of natural selection in optimizing the outcome of a fight between two animals." (Maynard Smith, 1972, p.26.) Whether matters are quite this simple will have to be discussed later. Certainly however, for the American sociobiologists, human aggression is just an extension of animal aggression.

4.2. SEX

We come now to the questions of sex and of mating. Once again, as with aggression, we go straight back to Darwin. Indeed, although Darwin first developed his theory of sexual selection in the animal world, it was not long before he was extending the theory and using it to understand humans. In fact, Darwin's classic on our species, the *Descent of Man*, devotes over half its pages to a technical analysis of sexual selection, which is then in turn applied to *Homo sapiens*: Darwin arguing both that racial differences and that sexual differences can be explained in terms of the mechanism. Given that sexual selection was divided into male combat and female choice, as one can well imagine the conclusions that Darwin drew harmonized nicely with some pretty conventional Victorian ideas about sex roles. "Man is more courageous, pugnacious, and energetic than woman, and has a more inventive genius." By way of compensation, woman has "greater tenderness and less selfishness". (Darwin, 1871, 2, pp.316, 326.)

In the sense that they too believe that theories about sex developed for animals can be applied directly to humans, the sociobiologists are very orthodox Darwinians. As we saw, sex and matings were seen, not as cooperative attempts to further the good of the species, but 'conflicts' where each partner tries to interact with the other in such a way as to further its own individual reproductive success. Moreover, because of the initial differential in sex-cell size and the fact that the female usually carries the fertilized egg for some time further, usually the 'battle' between the sexes was seen as one of males trying to get away with as much as they could, (*i.e.* trying to invest as little in the offspring as possible), and females trying to stop them or otherwise get compensation.[1] Because the human species so neatly fits the initial conditions, the sociobiologists have expected human sexual behaviour broadly to fit the general pattern, and they feel that they have not been disappointed. Indeed, even Dawkins, although he explicitly qualifies himself by stating that "man's way of life is largely determined by culture rather than by genes", feels compelled to conclude his discussion of sex by remarking on the extent to which human sexual behaviour follows general sociobiological patterns. ". . . it is still possible that human males in general have a tendency towards promiscuity, and females a tendency towards monogamy, as we would predict on evolutionary grounds." (Dawkins, 1976, p.177.)

So, starting at the beginning with Trivers' notion of parental investment, we have human females who, if impregnated, face the prospect of a nine-month pregnancy and a fifteen year stint of child-rearing if they are to bring their zygote to maturity, and we have human males who can certainly get someone pregnant with the minimum of effort (physiologically speaking), but who also have a stake in seeing that their offspring, who are born incredibly helpless and demanding, reach maturity. Since the male can impregnate with such ease, that is with so little effort, the first thing that one would predict is that males will have a tendency to impregnate any spare females that they can. But also, since human offspring require so much effort in raising, one predicts that males will try to get someone else to do their work for them! That is to say, there will be a selective pressure towards adultery (that is, males impregnating females who have mates to care for their offspring) and equally there will be strong selective pressure against being a cuckold. And Trivers is happy to note that anthropological data bear out this prediction, for "human male aggression toward real or suspected adulterers is often extreme". (Trivers, 1972, p.149.) Indeed, it has been shown that "when the cause is known, the major cause of fatal Bushman fights is adultery or sus-pected adultery. In fact, limited data on other hunter-gathering groups (in-cluding Eskimos and Australian aborigines) indicate that, while fighting is relatively rare (in that organized intergroup aggression is infrequent), the 'murder rate' may be relatively high. On examination, the murderer and his victim are usually a husband and his wife's real or suspected lover". (*Ibid.*)

But, intending to come back in a moment to the males, what about sex from the female perspective? After impregnation, she is more-or-less stuck with the baby — at least she is if she wants reproductively mature descend-ents. General sociobiological theory suggests that she has two, not necessarily incompatible strategies: what Dawkins calls the 'domestic-bliss' and 'he-man' strategies. First, she can try to pull the man into providing some parental care. She can do this by offering the male attractions before insemination, but by not being too ready to breed (what, in general terms, the sociobio-logists call 'coyness'). Trivers suggests that what we might find here is that human females rather adjust their strategies: attractive females have to offer less than unattractive females to attract males; hence less attractive females may in compensation raise the sexual sampling they are prepared to permit. (Trivers, 1972, p.146.) Perhaps also one might argue at this point that as the male starts to invest time, his interest in a not-too-rapid insemination starts to rise. If he is going to be committed to child care, then he wants to make sure that it is his own child for whom he is caring.

Wilson also rather suggests, no doubt connected with the fact that female humans need help for a very long time, that there has been selective pressure towards keeping the male drawn towards the female — thus, for instance, the female is sexually receptive at all times, and not just at the moment before or during ovulation. (Wilson, 1975a, p.554.) Probably also, because there is a strong selective pressure against adultery, it is not in the female's reproductive self-interest to get impregnated by stray males. Hence, the male is directed back to his mate, who expects some help in child care in return for sexual favours. Of course, none of this is to deny that the male has some direct in-terest in child care, given the helpless nature of the human infant — although probably male care and human helplessness are related, part cause and part effect.

But this is just one aspect of female sexual activity. The female not only wants help with the children, but also wants to bind her genes with 'good' genes from her mate. More precisely, it is in the females' reproductive interests not only to have help with child-rearing but also to bind her genes with genes from a mate, which mate's genes will, in turn, raise the offsprings' reproductive chances. At the moment, let us leave questions of consciousness out of the discussion, although obviously in the human case they do arise and will have to be tackled later. With respect to this question of 'good' genes, it is clear that sociobiologists see something akin to Dawkin's he-man strategy at work here, with it no doubt being responsible for human sexual dimorphisms. Trivers, for example, suggests that, even in contemporary society, if they can, females choose 'he-men' – that is, people further up the socioeconomic scale! (Trivers, 1973, p.172.) Also human polygamy, far more common as a husband with wives than vice-versa, may be involved here. As in animals, females mating with an already-mated male at least know that they are going with a 'winner', perhaps both in the matter of care available and in the matter of genes.

As pointed out in the case of animals, thinking purely in terms of female choice at this point is a little misleading. What one has almost invariably is males competing, perhaps for the best females (in which case one has male choice also) and females allowing themselves to be chosen and perhaps in turn also choosing. It is clear it is basically this overall picture that sociobiologists want to read into human sexual activity and relations. Although, unlike some earlier authors, sociobiologists have not gone into the precise details as to how male sexual characteristics have evolved, they see males as being more aggressive and stronger than females, doing their best to impress their wills on females. Trivers implies that it is the more successful males who get the choice of the females, although conversely also one has the more attractive females choosing the more successful males. (Trivers, 1972, p.172.) Wilson also quotes favourably, or at least without negative comment, ideas suggesting that important in human evolution has been male combat and display towards females. Moreover, he sees today's society as though rather fashioned by these sorts of things – for instance, he implies that generally males are dominant over females. (Wilson, 1975a, p.552.)

Of course, as we know, 'good' genes are a relative thing: what is good for one animal is not necessarily good for another. In particular, there may well be an advantage in promoting heterozygosity and thus breeding with forms not too similar to ones own; even in the animal world there is some evidence to back up this supposition. Similar reasoning is applied by sociobiologists to the human case. Specifically, they point to the almost-universality of incest taboos, and they suggest that these might have a genetic foundation, having been fashioned through selection. And in support, the sociobiologists show that such taboos certainly make very good biological sense. The effects of close inbreeding amongst humans – brother–sister matings for instance – are absolutely horrendous. Without a doubt, any genes promoting incest taboos will be strongly favoured by selection. Also, given incest taboos, the probabilities of local mate competition are reduced: brothers will not compete sexually for their sisters. (Alexander, 1974; 1977b.) Moreover, notes Alexander, the phenomenon of falling in love seems just what would be expected given the truth of sociobiology. Two people, often strangers, meet, and suddenly, 'without reason', they feel a strong sexual attraction towards each other. No more perfect mechanism for promoting out-breeding could be imagined. (Alexander, 1977a.)

Before leaving this section, it will be useful to make one final comment here. The basic problem at issue, or at least one of the basic problems at issue, obviously, is the extent to which human sexual patterns can be said to be a direct function of the genes, rather than of essentially non-genetic culture. It would be premature here to get into detailed discussion of the truth of the sociobiological claims, for the matter must be raised at some length later. But, in presenting the sociobiologists' case it should in fairness be added that they argue, not only from the supposed fit of their theory with the facts, but also analogically. It is suggested that in non-humans, where cultural non-genetic factors can almost definitely be excluded, sexual patterns in important respects similar to those of humans obtain. Moreover, the sociobiologists argue that in some cases it is legitimate to argue from non-humans to humans. Hence, we have analogical support for the genetic basis of human behaviour. Thus, for instance, Wilson argues analogically for the genetic basis of human male dominance from the fact that such dominance is to be found fairly generally through the non-human primates, and the fact that he thinks the links between us and them are strong enough to bear such a conclusion. I shall say no more here; but this dual approach to confirmation, direct and indirect, should be remembered.

4.3. PARENTHOOD

Again, in this section, it is to the work and ideas of Trivers that we must turn primarily, and as always it will be remembered, the dominant theme is one of reproductive 'self-interest'. (Trivers, 1974.) What strategy will best maximize my genes in future generations? Whether done consciously or not, the sociobiological claim is that all such strategies are essentially under the control of the genes.[2]

Even at the point of birth, if not conception, the question of reproductive interest is a matter of importance, because as we have seen sex-ratios are a function of maximum future gene representation. Normally, the ratio favourable to the parent is 50:50, at least in effort that must be invested, and it is clear that, in this respect, humans are fairly normal. However, we know that intervening factors can affect sex-ratios. One such possible factor is local mate competition. Another suggested factor is based on maternal physical fitness (*i.e.* health). In particular, because females in many animal species are the limiting resource, in such species females are far more likely to have offspring than males. Hence, because unhealthy males will probably be more severely handicapped than unhealthy females, an unhealthy mother likely to raise unhealthy children would be best advised (evolutionarily speaking) to have daughters, for they will be more likely to reproduce than sons. Trivers and Willard (1973) have found that this conclusion holds, not only in the animal world, but in humans also. In particular, females under adverse conditions tend to have fewer children but proportionately more daughters. The favourable implications of this finding for the views about males and females expressed in the last section will, no doubt, not escape the reader.

But what about when one has had the child and one has made the commitment to raise it? Sociobiological theory suggests that conflict will arise because the interests of parent and child are only partially the same: in some respects they are genetic rivals, particularly when benefits from parent to child fall into the twilight zone where the parent would be better off investing in another child, but the first child receives more from the investment than it would if the investment went to the other child (roughly, it must do a full-sibling at least twice as much good as it would do the child itself). There seems no doubt to sociobiologists, to Trivers and Wilson at least, that one can legitimately extend this view of parent—child relations to humans. Conflicts between parents and children over parental help are direct functions of their biologies — there comes a time during growing up when children want more than it is in the interests of parents to give. And, of course, generally, the parent is trying to minimize the investment it has to make and (within limits) the child to maximize the investment it gets.

Even that all-too-familiar time at the end of the day is included here. "For example, when parent and child disagree over when the child should go to sleep, one expects in general the parent to favour early bedtime, since the parent anticipates that this will decrease the offspring's demands on parental resources the following day." (Trivers, 1974, p.260.) Moreover, suggests Trivers, through this theory of parent—offspring conflict, one can explain some interesting and otherwise puzzling psychological phenomena. For instance, it may well be that willingness to provide human parental care is triggered by certain behaviour by the offspring (as, for example, begging in birds). If this is so, then regressive behaviour could in some way be an attempt to elicit more attention and care from parents. ". . . at any stage of ontogeny in which the offspring is in conflict with its parents, one appropriate tactic may be to revert to the gestures and actions of an earlier stage of development in order to induce the investment that would then have been forthcoming . . . A detailed functional analysis of regression could be based on the theory presented here." (Trivers, 1974, p.257.)

In the case of humans, Trivers feels emboldened to push his analysis of parent–offspring conflict right to the point where it becomes virtually a species of parental manipulation. Altruistic behaviour by an offspring might benefit a parent more than the offspring itself. Thus, when an individual helps a first cousin it is helping an individual related only 12.5% to it. However, it is helping an individual related 25% to the parent. Hence, it may well be in the parent's interest to push the offspring towards altruism, but not the offspring's to comply. Even with total strangers, one might get such a conflict over altruism. If the stranger is going to respond not just to the individual but to the individual's siblings too, the benefit to the individual might not be great enough to spark altruism, but the benefit to all of the siblings might be great enough to make the parent push the individual towards altruism.

In short: ". . . a fundamental conflict is expected during socialization over the altruistic and egoistic impulses of the offspring. Parents are expected to socialize their offspring to act more altruistically and less egoistically than the offspring would naturally act, and the offspring are expected to resist such socialization." (Trivers, 1974, p.210.) In other words, causally speaking, teaching children things like honesty, decency, generosity, is not so much (or at least not exclusively) a question of introducing children to culture, in a general sense, but rather of driving the children to support the parents' biological interests. "Conflict during socialization need not be viewed solely as conflict between the culture of the parent and the biology of the child; it can also be viewed as conflict between the biology of the parent and the biology of the child." (Trivers, 1974, p.160.)

Because we are now obviously starting to examine the sociobiological explanations of human altruism, let us acknowledge this fact explicitly, and using the same structure as in the last chapter, consider, in turn, the three suggested mechanisms for such altruism: kin selection, parental manipulation, and reciprocal altruism.

4.4. KIN SELECTION

It is easy to see why sociobiologists keen to apply their theories to the human dimension would look upon kin selection with pleasure. Any help given to relatives is grist to its mill. Of course, it might be pointed out that we, in Western society, tend not to live that close to relatives, other than the very nearest (*i.e.* those in the nuclear family), so by definition, at this point, kin selection cannot be used to cover that many cases of human interaction, altruistic or otherwise. However, the sociobiologists argue that one cannot fairly take Western society as a paradigm. They do not deny that vast cultural changes can and do occur, lying across our biological nature and altering and hiding it — although the sociobiologists would deny that this nature is as altered or hidden as many of us are wont to think. But the point is, as Wilson (1977a) says, our cultures are "jerry-built on the Pleistocene". Where something like kin selection would be or have been really important would be in less technologically advanced societies today and in the past.

And here the sociobiologists feel that they are on fairly firm ground in arguing that kin selection has been an important molding factor in human nature. A great many recorded pre-literate societies show an almost fanatical obsession with kin, working out relationships in the most minute detail. "Man is aware, to an extraordinary degree, of differences in his relationship to the other men with whom he lives." (Alexander, 1971, p.117.) Moreover, people in the pre-literate societies let their social behaviour be governed by these kinship relationships: roughly speaking, the closer the relatives the more you are prepared to do for them. This, of course, is precisely what one would expect if kin selection had been an important causal factor in the development of the human genotype. Furthermore, argue the sociobiologists, there is good reason to believe that kin selection may have been important in the past. As was seen in our discussion of aggression, Alexander particularly has argued that kin selection was crucial in the development of human intelligence. His hypothesis "calls for war to be waged in some relationship to degrees of genetic difference and raises the question of the selective value and background of assisting one's closer relatives at the expense of non-relatives or distant relatives. In any species engaging in the more violent kinds of intraspecific competition, ability to recognize and spare close relatives would be highly favoured". (Alexander, 1971, p.117.) Moreover, adds Alexander: "In some modern hunting-gathering peoples still living in small groups, and in which intertribal aggression is prevalent, the extent and nature of such knowledge [of kin] has amazed anthropologists more than any other of their attributes." (*Ibid.*)

Of course, the critic might object that sometimes kinship systems do not work – at least, they do not work in the way that is predicted by sociobiology. However, the position of the sociobiologists is that any conflict between their predictions and reality tends to be apparent rather than real: indeed, they rather suggest that conflict often points the way to sociobiological triumphs. Thus, for example, take the mother's brother phenomenon. In some societies, responsibility for the children falls on the mother's brother rather than (as kin selection would seem to predict) the father. Surely, we have a conflict here with males supporting 25% relatives rather than 50% relatives? However, Alexander (1977a) suggests that if it turns out – as there seems to be evidence that it does – that in such societies paternity is frequently in doubt (whatever the social father) then mother's brother's care is precisely what kin selection would predict. One is passing up a very dubious 50% relationship for a very certain 25% relationship (presumably somewhat less if siblings are liable only to have a common mother).

To summarize, whenever general living conditions or other society-wide circumstances lead to a general lowering of confidence of paternity a man's sister's offspring, alone among all possible nephews and nieces, can become his closest relatives in the next generation. In consequence, so long as adult brothers and sisters tend to remain in sufficient social proximity that men are capable of assisting their sister's offspring, a general society-wide lowering of confidence of paternity is predicted on grounds of kin selection to lead to a society-wide prominence, or institutionalization, of mother's brother as an appropriate male dispenser of parental benefits.

This appears to be exactly what happens. The evidence indicates that lowered confidence of paternity, fragility of marriage bonds matrilineality, and shifts toward a prominence of mother's brother go together, in a pattern almost dramatically consistent with a Darwinian model of human sociality. (Alexander, 1977a, p.17.)

Another example cited by Alexander as an anthropological phenomenon apparently falsifying sociobiology, but in fact triumphantly supporting it, involves cousins. Many societies make a distinction between 'parallel-cousins', offspring of siblings of the same sex, and 'cross-cousins', offspring of siblings of the opposite sex. Moreover, such societies often treat parallel-cousins as being much closer relatives than cross-cousins, even on occasion referring to them as 'siblings'. *Prima facie* this contradicts sociobiological theory, because the relationship between parallel-cousins and cross-cousins ought to be the same, namely $\frac{1}{8}$. However, argues Alexander, what we find, in fact, is that there is a significant correlation between societies which make such a cousin distinction, treating parallel-cousins as closer than cross-cousins, and societies where indeed parallel-cousins tend to be closer biological relatives than cross-cousins: this latter closeness coming through such facts as that one is dealing with polygynous societies where males have more than one wife, often sisters, and hence parallel-cousins are frequently half-siblings also. Thus, sociobiological theory is supported, for "asymmetrical treatment of cousins is very strongly concentrated in precisely the kinds of polygynous societies in which it is predicted to occur from an inclusive fitness model, and least frequent in the kind of monogamous society where that is expected". (Alexander, 1977a, p.18.)

Before leaving kin selection, it might be appropriate to return again to human societies in general, including our own. Showing the way in which sociobiologists think that something like kin selection can have implications even for societies which seem furthest from their biology, the suggestion has been made that kin selection may well throw light on a phenomenon which apparently affects a sizeable minority of any society: homosexuality. From a biological point of view, homosexuality presents something of a puzzle. Why would people have homosexual inclinations, surely directing them away from reproduction? If there is any evidence that homosexuality might, in part, be controlled by the genes, and the sociobiologists think that there is, then would not selection have wiped it out pretty quickly? And yet apparently anywhere up to 10% of human beings have fairly significant homosexual leanings.

One suggestion which has been put forward is that homosexuality may be a product of balanced heterozygote fitness. (Wilson, 1975a, p.555.) Heterozygotes carrying one 'homosexual' gene are, biologically speaking, super-fit, and this balances out homosexuals, who are the biologically less fit phenotypes of genotypes homozygous for 'homosexual' genes. (The genes themselves are not homosexual! What is meant is that such genes can cause homosexuality in the phenotype.) Another suggestion which has been made, however, is that homosexuality may be a product of kin selection. (*Ibid.*) Homosexual genes may have lowered personal fitness, but in freeing one from the need to support one's own mate and children, one was therefore better able to help one's own close kin to survive and reproduce. In other words, homosexuals increase their inclusive fitness through their homosexuality: they are rather like sterile worker ants. Therefore, even though homosexuals today may not be intimately concerned with raising kin, they have their genes and their consequent personal sexual desires because of kin selection in their pasts.

Another possible phenomenon shaped by kin selection may be the menopause. It is often remarked how odd it is that females lose reproductive ability rapidly, whereas males tail away gradually. Given the fact that women are stuck with most of the child-rearing (for reasons discussed above), there may be an advantage to losing personal fertility. Obviously, it is in one's own reproductive interests to raise children than grandchildren (50% to 25%); but, if as old-age approaches one's chance of raising a child to maturity fades, then it might raise one's inclusive fitness if one spends one's effort helping with the grandchildren, at a time when such help is most needed, and where the final year's of rearing effort can be done by the unaided still-living parents. Hence, kin selection promotes altruism towards grandchildren by bringing on the menopause.

4.5. PARENTAL MANIPULATION

We turn next to parental manipulation, and indeed we have seen already, in Trivers' analysis of parent–offspring conflict, suggestions that parents try to manipulate their children into altruistic behaviour. Remember the difference between this and kin selection. Although the altruism that the parents are trying to impress on their child will in fact benefit the children's kin, the main genetic recipients of the altruism will be the parents themselves. It is they who have the most to gain: not their reluctantly altruistic children. Of course, in the cases we have been discussing in the last section, everything comes out alright in the end; at least, presumably later the parents are going to force their younger children into altruism towards their first-born, even though these younger children might not like it. However, it is suggested that sometimes parental manipulation goes further than this — to reducing or suppressing permanently a child's interests in favour of siblings.

We know already that this extreme form of parental manipulation is possible genetically: suppose that a parent has five children, that without aid two will reproduce, but that with aid three out of four will reproduce. Hence, if the parent carries genes making it, perhaps subconsciously, mould one child into a non-reproductive altruist towards the others, such genes will be favoured by selection, even though they may not act entirely in the altruist's self-interest. One suggestion where one might get an extreme form of parental manipulation has come from Alexander (1974). Just as in certain species of insect parents lay so-called 'trophic' eggs, that is eggs which are not allowed to develop but instead are used as food by other offspring, Alexander surmises that we might have similar behaviour for similar reasons in humans. For example, in times of food-shortage it is not unknown for aborigines to

feed their youngest child to their older children: the youngest child is being
sacrificed for the sake of the older children, and clearly, from a biological
viewpoint, this sacrifice is in the reproductive interests of the parents.

At this point, the reader might be feeling a little uneasy. Technically
speaking, if one understands by 'altruism' giving up or appearing to give up
one's reproductive interests for the sake of another, such 'trophic' children
may indeed qualify as (unwilling) altruists. But it is obvious that in the human
case the sociobiologists want to go somewhat further. Clearly, they want to
capture the causal content of the emotions and behaviours that we humans
have and describe, even without any knowledge of biology. They want to
explain biologically what we understand generally by, say, 'aggression' and
'altruism', where this latter in some way means making an effort for others:
"Regard for others as a principle of action." Now, in the next chapters, we
shall have to examine in some detail the extent to which what the sociobiologists purport to prove with their technical notions coincides with our
non-technical beliefs, findings, and so forth. However, one might with some
justification feel that this instance of trophic children shows already that
there cannot be a perfect mesh between the sociobiological claims and our
non-technical understanding, simply because it is just not appropriate in ordinary language to speak of a child fed to its siblings as an 'altruist'. Altruism in
the human case perhaps demands some sort of intention by the altruist, and
this the child does not have.

Perhaps, at this point, the best thing to do is just to take note of this
uneasiness (my uneasiness!) and turn to some other cases of parental manipulation, involving human offspring beyond the infant stage, where at least the
possibility exists that the genes might be causing behaviour which corresponds
to what we in general discourse might call 'altruism' (without at this point
categorically claiming that we would indeed call such behaviour 'altruistic').
One possible such instance revolves around the fact that, in some countries, a
family's livelihood is centred on a farm: if this goes out of the family or is
divided up, then the family suffers. Alexander (1974) suggests that in such
situations one often gets substantial parental manipulation, forcing the
younger sons not to seize a share of the property but in some way to behave
helpfully to the oldest brother, who takes title to the farm. This help might
take various forms, not just direct physical help. The younger sons might be
steered towards non-reproductive life-styles, for example to an altruistic
priesthood. Perhaps one special form of this manipulation is to be found in
polyandrous societies, that is societies where wives have more than one husband (who are usually brothers). The oldest son has title to the farm and the

children; but, in return for help rendered, the younger brothers are offered sexual favours by the joint wife: the brothers being forced into such a situation by the parents. Interestingly, richer men in such societies tend to monogamous or polygynous (husband with wives).

It will not have escaped the discerning reader that we have in parental manipulation another possible cause of homosexuality. It has certainly not escaped the discerning sociobiologist. (Trivers, 1974.) Parents manipulate some of their offspring into altruistic but non-reproductive desires, and one of the surest and safest ways to do this is to make such offspring homosexually inclined. Notice that in this kind of case there are not genes for homosexuality as there may be in the kin-selective case (there are, of course, genes which will make one homosexual under certain environmental conditions, but unlike the kin-selective case it is possible that everyone has these). Notice also, that it is not being implied that parents consciously manipulate children into homosexuality. Indeed, from the genes' viewpoint, things might be a lot more efficient if everyone, parents and children, is ignorant of what is going on.

And, finally, even though parents might not make homosexuals out of their children, they may well try to manipulate the children into unions which are not necessarily to the children's best interests. "Since in humans an individual's choice of mate may affect his or her ability to render altruistic behaviour toward relatives, mate choice is not expected to be a matter of indifference to the parents. Parents are expected to encourage their offspring to choose a mate that will enlarge the offspring's altruism toward kin." (Trivers, 1974, p.261.) Hence, there may be pressure to marry cousins or to forge alliances with other groups, and certainly to avoid unions with society's pariahs: such unions do no one any good, least of all the parents and their kin.

4.6. RECIPROCAL ALTRUISM

Finally, we have reciprocal altruism. Trivers (1971), in particular, has been concerned to analyse this in the human context. He suggests that all human societies show signs of altruistic behaviour (quite apart from altruism to relatives): helping the old and sick, helping those in danger, and so on. Furthermore, he feels that humans with their various attributes, long life span and the like, would have been just the sorts of beings to have had this behaviour promoted by the genes. Hence, he argues: "There is no direct evidence regarding the degree of reciprocal altruism practiced during human evolution nor its genetic basis today, but given the universal and nearly daily practice of reciprocal altruism among humans today, it is reasonable to assume that it has been an important factor in recent human evolution and that the underlying emotional dispositions affecting altruistic behaviour have important genetic components." (Trivers, 1971, p.48.)

Having made this assumption, Trivers then feels he is in a strong position to draw conclusions about human psychology, all of which seem to have independent empirical backing. A number of these conclusions are as follows: First, we should not expect people to be complete altruists. If they can cheat and get away with it they should; and certainly just about everybody should in some respects be on the verge of cheating and, conversely, very sensitive to the possibility of cheating in others. Second, people should, under normal circumstances, be more generous towards friends and less generous towards enemies. In other words, we should be more ready to help those who help us and less ready to help those who do not help us. Third, we should be ready to jump on cheaters. "Once strong positive emotions have evolved to motivate altruistic behaviour, the altruist is in a vulnerable position because cheaters will be selected to take advantage of the altruist's positive emotions. This, in turn, sets up a selection pressure for a protective mechanism." (Trivers, 1971, p.49.) And this Trivers finds in the strong moral tone people take towards cheating. Fourth, people should be sensitive to the need and cost of altruism. The more you put out, relative to what you have, the more credit you should get (*i.e.* the widow's mite). Conversely, the more you need help, the more ready people should be to respond. ". . . crudely put, the greater the potential benefit to the recipient, the greater the sympathy and the more likely the altruistic gesture, even to strange or disliked individuals." (*Ibid.*) Fifth, there should be selection towards guilt and a willingness to make reparation. If a cheater has been caught out, then it is both in his or her interest to become accepted again as trustworthy, and in the interest of the cheated to have the cheater turn trustworthy. Guilt and making amends help both cheater and cheated to return to a position of trust. Of course, if the cheater is not found out, one should get nothing like the guilt and repairing actions. Sixth, linked with some of the previous points: "once friendship, moralistic aggression, guilt, sympathy, and gratitude have evolved to regulate the altruistic system, selection will favour mimicking these traits in order to influence the behaviour of others to one's own advantage." (Trivers, 1971, p.50.) In other words, selection doth make hypocrites of us all, or, at least, a good many of us. Seventh, linked with the previous point, selection will make us out to spot hypocrites! When people show too little, or too much, concern, others will start to suspect their motives and their sincerity. Eighth, people will be ready to initiate or cement relationships, because reciprocal altruism generally is a good thing. Hence, in certain circumstances one might be more generous towards strangers or enemies than towards friends. Ninth, in the human situation, we should expect multiparty interactions: it may, for example, pay people to gang up on cheaters, and to repay debts to altruists' kin. Tenth, we should find plasticity. "For example, developmental plasticity may allow the growing organism's sense of guilt to be educated, perhaps partly by kin, so as to permit those forms of cheating that local conditions make adaptive and to discourage those with more dangerous consequences. One would not expect any simple system regulating the development of altruistic behaviour." (Trivers, 1971, p.53.)

All of these predictions, Trivers believes, can be substantiated. Some facts perhaps can be more readily seen in pre-literate societies; others in our own society. But, by and large, they are human constants, many of them are not very obvious, and no other theory than biological reciprocal altruism – certainly no purely psychological theory – can explain them all. Hence, the case for the importance of human reciprocal altruism, promoting genes which control the human altruistic system, is strong.

4.7. A GENERAL MODEL FOR HUMAN ALTRUISM

We have now covered the sociobiologists' treatments of the various mechanisms for the promotion of human altruism – at least we have covered the treatments of the mechanisms taken separately. In concluding this expository discussion of human altruism, and indeed in concluding our discussion of human sociobiology generally, we might just note that Alexander (1975) has tried to combine some of the mechanisms for altruism into a single model. In particular, referring to a purely non-biological analysis of pre-literate societies by the anthropologist Marshall Sahlins (1965), Alexander suggests that Sahlins' findings can best be interpreted by sociobiological theory. Sahlins is concerned with the setting up of a general model of reciprocity in pre-literate groups: how do people respond to others? In a roughly circular, radiating fashion, Sahlins divides interpersonal relationships into three categories: they show generalized, balanced, and negative reciprocity. (See Figure 4.1.)

So, first, we have generalized reciprocity, practiced basically amongst kin, that is amongst relatives. "A good pragmatic indication of generalized reciprocity is a sustained one-way flow. Failure to reciprocate does not cause the giver of stuff to stop giving: the goods move one way, in favour of the have-not, for a very long period." (Alexander 1974, p.91.) Fairly obviously,

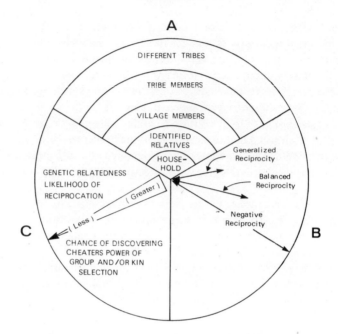

Fig. 4.1. (From Alexander 1975 and Barash 1977.) We have here a correlation between (A) Social Relationships, (B) Patterns of Reciprocity, and (C) Factors Influencing Altruism in Preliterate Cultures. Sahlins (1965) identified the first two phenomena and Alexander (1975) claims that they can be explained by (C).

Alexander immediately identifies generalized reciprocity as a function of kin selection. Next, we have balanced reciprocity. This tends to be something which takes place outside of the family, but within the tribe: in other words, it takes place between unrelated friends and acquaintances. The chief difference between generalized and balanced reciprocity is that, in the case of the latter, you do expect a return: something which balances one's own gift. "In precise balance, the reciprocation is the customary equivalent of the things received and is without delay . . ." (Alexander, 1975, p.92, quoting Sahlins, 1965.) Putting the matter bluntly, if one does not get returns for what one lays out, one ceases to lay out. And as might be expected, equally immediately, Alexander identifies balanced reciprocity as a function of reciprocal altruism.

Finally, we have negative reciprocity. This occurs outside the tribe or nation, because it involves dealing with other tribes or nations. It brings into play all kinds of actions which will get you something for as little as possible, whether one has to cheat, lie, steal, fight, or what. Anything goes: no holds are barred. Of course, Alexander is just as delighted with this kind of reciprocity as with the others, for we are now at the point where the mechanisms for altruism break down, and if anything we start to verge into the aggressive range, which is obviously just about where negative reciprocity lies: half-way between altruism and aggression, and perhaps a little closer to the latter. In short, given the exactness between his sociobiological expectations and Sahlins' independent anthropological taxonomy, Alexander feels that the genetic approach to humans scores a significant victory.

And so now, with Alexander's sociobiological interpretation of Sahlins' anthropological model, we conclude our look at human sociobiology, and with it the first, expository part of this book. The time has come to let the critics have their say.

NOTES TO CHAPTER 4

[1] Please note that 'conflict' and 'battle' here refer to organisms having different reproductive interests. In fact, as we shall learn, behaviourally we might get harmonious cooperation much of the time, especially in a case like humans where so much parental care is required.

[2] Of course, a major difference between humans and animals, at least most animals, is that humans have consciousness and reasoning powers, and animals do not. In a sense, however, consciousness is not of primary importance to human sociobiologists. Their claim is that whether consciousness is involved or not, humans (like animals) act so as to maximize their fitness. In fact, as we shall be able to infer, the belief is that for humans sometimes consciousness is involved, and sometimes it is not — but its existence and involvement are secondary, and thought not vital to a first-order discussion of human behaviour. What count are actions, however caused. Obviously, a position like this implies that much of the language of sociobiology (animal and human) expressed as it is in such terms as organisms 'wanting' to maximize fitness is at best metaphorical, because these are terms literally applicable only to consciously acting beings. Later, I shall discuss this question of metaphor.

CHAPTER 5

NORMATIVE CRITICISMS

We saw, in the first chapter, that the main critics of human sociobiology, certainly the most vociferous and vitriolic, have been the members of the Cambridge-based Sociobiology Study Group of Science for the People. (Allen *et al.*, 1976, 1977.) However, we know also that, since the initial outbursts, more temperate critics have appeared, particularly − of all people − the anthropologist Marshall Sahlins (1976). No doubt, nothing stirs one to action quite as much as having one's ideas taken to support conclusions one abhors. In this and the next chapter, I shall run through the various criticisms which have been levelled against human sociobiology: my concern in this chapter will be with criticisms which ultimately raise questions to do with values, and my concern in the next chapter will be with criticisms aimed at showing that in some way human sociobiology fails as genuine science. Because I shall have, with the criticisms, given both sides of the matter, I shall feel free to comment myself as we go along. I leave until the following chapter, however, an overall assessment of the strength of human sociobiology.

5.1. SOCIOBIOLOGY AS REACTIONARY

The Sociobiology Study Group of Science for the People, the 'Boston critics' for short, start their critique of sociobiology by placing it firmly in the tradition of biological determinism, a science-cum-philosophy which "attempts to show that the present states of human societies are the result of biological forces and the biological 'nature' of the human species". (Allen *et al.*, 1977, p.1.) This, of course, seems fairly indisputable, although one might be forgiven for wondering how any modern scientific theory of humans could avoid being in some sense biologically deterministic, even the most environmentalist, given the way that biologists perpetually emphasize how organisms are a product both of their genes and of their environment. But, obviously, in a sense this is a quibble: what the critics want to do is to point to the fact that sociobiology relates humans firmly to their genes in a way that environmentalists do not. And, in pointing this way, they are clearly right: sociobiology does nothing, if it does not do this.

It goes without saying that the critics have more in mind than merely a taxonomy of biological disciplines. Such determinist theories, we learn, reflect 'socio-economic prejudices', they are apologetics for the *status quo*; and indeed they lead to, certainly support, the philosophy of the gas-chamber.

For more than a century, the idea that human social behaviour is determined by evolutionary imperatives and constrained by innate or inherited predispositions has been advanced as an ostensible justification for particular social policies. Determinist theories have been seized upon and widely entertained not so much for their alleged correspondence to reality, but for their more obvious political value, their value as a kind of social excuse for what exists. (Allen *et al.*, 1977, p.3. See also Allen *et al.*, 1976, p.182.)

And the critics flesh out their case by citing the instance of Konrad Lorenz, who began with geese and ended by echoing Hitler's filthy racial policies of extermination.

Taken at its most immediate level, this criticism strikes me as being intemperate to the point of unfairness — cruel even. Indeed, it so overstates the case that there is a danger that, in emotional sympathy for the sociobiologists, one might glide over points at which they might be open to genuine criticism. Let it be stated categorically: the sociobiologists are not racists. There is no suggestion at all in their writings that (for instance) blacks are inferior because of their genes. Even less is there suggestion that we might properly embark on wholesale eugenics programmes to eliminate genes of certain racial types, like blacks or Jews. I do not know the political views of the sociobiologists, but they are certainly not neo-Nazis.

With the air now cleared, let us now turn to more moderate versions of the charge that there is something politically reactionary, even dangerous, about sociobiological writings. For a start, it might be agreed that there is nothing intentionally racist about sociobiologists and their writings, but, nevertheless, it might be feared that there are elements in what they produce which could be taken up by less-than-honourable people and used to justify vile social doctrines. Genetic speculations about humans have led to such 'justifications' in the past: perhaps they could do so again. Hence, simply because sociobiology is a species of human genetics — particularly human behavioural genetics — it is dangerous.

In reply, let us first make the obvious point that no one is going to deny — least of all eminent biologists as are included in the critics — that humans are, in essential respects, functions of their genes, and moreover that there are genetic differences between people, and more specifically between women and men of different races. Blacks are not black and whites are not white simply because there is more sunshine in the Congo than in Canada (although this is not to deny that the sun might have been a factor in skin colour selection). Moreover, simply drawing attention to the fact that there are genetic differences between people seems in itself not particularly racist or leading to racism, nor does an examination of the way in which different genes affect people or the possible adaptive significance that genes have or had.

Of course, the critics might reply at once that this is precisely to miss the point. As soon as one starts talking about and studying genetic differences, any genetic differences, this opens the way to racism. Even if genetic differences really exist, as they obviously do, drawing attention to them is the first step to discrimination on the basis of them. Perhaps Jews really do have big noses, perhaps they do not; but this is not to deny that one is going to be very suspicious of any research-grant proposal which intends to do a comparative study of Jewish—Gentile nose sizes. However, although one would have to be naïve not to say insensitive to the history of this century not to allow at least some force to this charge, there are a number of powerful objections to it as a foundation of a programme of action, or non-action.

First, there is what seems to be the general danger of proscribing an area of research. (Ruse, 1978c.) It may be unfashionable to say so, but one of the glorious things about human beings is the way in which they inquire into their world, whether it be through science, or literature, or philosophy, or whatever. In itself, I believe free inquiry to be a good, and denying such inquiry is *prima facie* wrong. Obviously, at the centre of the objections to sociobiology is the belief that humans are more than just animals (otherwise why all the fuss?), and so the critics themselves should be sympathetic to this point I am making.

Of course, one might reply that although, in itself, it is a general good, there are times when free inquiry must be barred — for instance, when it poses too great a cost or threat to the community at large — and that study of human genetics is one such case. (Jonas, 1976.) But this brings me to my second objection, namely that I do not see that the study of human genetics (including behavioural genetics) falls into this category. It is indeed true that some bad men and women in the past have misused human genetics, but this, in itself, is no argument for ignoring human genetics today. If it is such an argument, then we ought also to proscribe physics, and chemistry, and psychology, and many other disciplines, for they too have been misused in the past. Perhaps also we ought to proscribe philosophy, for this seems to have been the most dangerous subject of the lot!

Furthermore, although it is perhaps true that some bad people will even today misuse the findings of genetics (and physics and so on), this is no argument against studying human genetics. Science can be used for good as well as bad, and this holds particularly of human genetics (specifically including behavioural genetics). On the one hand, it can be used to dispel, as well as create, prejudice. Certainly, for instance, no one who knows anything at all about modern human genetics can for one moment suppose that Jews are in some way a 'degenerate' race. Apart from anything else, we know that there is too much genetic similarity between Jews and the rest of us to countenance such a judgment. Or, to take a case mentioned by Wilson, we know that there is no genetic foundation to caste difference in India. (Wilson, 1975a, p.555.)

On the other hand, human genetics might very well be used in a positive manner to help humankind. Take, for example, a suggestion by Wilson that xenophobia (fear of strangers) may have a genetic foundation: "Part of man's problem is that his intergroup responses are still crude and primitive, and inadequate for the extended relationships that civilization has thrust upon him ... Xenophobia becomes a political virtue." (Wilson, 1975a, p.575.) It is beyond question that one of the biggest problems we humans face today is the strife between peoples, whether it be between Catholics and Protestants in Northern Ireland, or, on a large scale, such as during the two world wars which erupted. Moreover, it is a problem which is intensified because of the horrendous weapons which we now have. Surely, any understanding of what makes people behave as they do when faced with strain or with strangers or the like, cannot but be a major step in coming to terms with the threat of holocaust which hangs over us all? It is true that Wilson's suggestions about xenophobia may prove groundless; but to deny study of this kind seems akin to denying that the search for the causes of cancer has any relevance to its cure. (Note: I am not, at this point, questioning the truth or falsity of sociobiological claims, or even whether they make very good sense, but whether they are in some sense inherently politically reactionary.)

I argue, therefore, that, in itself, the study of human genetics, including here human behavioural genetic theories like sociobiology, is not explicitly or implicitly racist and that there is no definitive case for proscribing it: indeed, the opposite holds.

At this point, an objection might be raised to my argument thus far. A critic might complain that unfairly I have been combining sociobiology with morally neutral studies of the genetic foundations of human phenotypes. Such a critic might readily concede that there is nothing racist or otherwise offensive about inquiries into, say, the reasons why human beings have different heights, including, here, genetic reasons. Such a critic might even allow some studies on possible genetic causes of human behaviour: were the Bachs brilliant musicians purely because of their environment or also because of their shared genes? What is offensive is a certain kind of human behavioural genetics, namely the kind which comes up with sweeping racial generalizations, such as that blacks are lazy or stupid, Jews are avaricious, Englishmen are emotionally cold, and the like. The trouble with sociobiology is that it belongs to this species of human behavioural genetics.

In reply to this objection, first it must be re-emphasized that the sociobiologists do *not* claim that certain identifiable races have certain identifiable genetically-caused bad or unfortunate habits, such as those just listed. It is indeed true that, at one point, Wilson writes: "Moderately high heritability has been documented in introversion–extroversion measures, personal tempo, psychomotor and sports activities, neuroticism, dominance, depression, and the tendency toward certain forms of mental illness such as schizophrenia . . . Even a small portion of this variance invested in population differences might predispose societies toward cultural differences." (Wilson, 1975a, p.550.) But aside from the fact that this is just untested speculation, Wilson certainly does not follow up by saying that all or even most people in some populations are likely to be crazy in some way.

Apart from anything else, we know that Wilson postulates a 'multiplier effect', and so if anything his position seems to be that if indeed there prove to be significant differences between populations, say with respect to numbers of schizophrenics, this does not mean that one culture as a whole will be more schizophrenic than another. The ultimate effects might be quite different. Joan of Arc is often labelled a schizophrenic; she certainly had a great impact on her countrymen; but the overall effect was hardly to make the French as a whole mentally deranged. Only the most extreme reductionist would argue that from the behaviour of a number of individuals in a population can we deduce the culture. Joan did not make Frenchmen more schizophrenic: right up to de Gaulle, who admired her greatly, she made the French more patriotic, more courageous, and so forth.

The one sociobiologist who has really tried to pin down the culture's of specific societies to genetic foundations is Alexander (1974). But, in his case also, unless one has the ethics and viewpoint of a nineteenth-century Christian missionary, there is no implication that certain peoples are bad because of their genes. Even the aborigines, supposedly feeding their babies to their older children, are hardly to be condemned. They are taking very drastic action, in the face of very drastic circumstances, precisely to save as many of their children as they can.

The second point to be made against this critic is that the kinds of data concerning racial differences used by the sociobiologists is, in fact, allowed by the critics! Indeed, one of the few firm claims Wilson makes as to differences is taken from Lewontin (1972), namely that human blood-type systems show that 85% of the variance is intra-population, and only 15% is inter-population. Now, one may want to question the conclusion that Wilson draws, namely that: "There is no a priori reason for supposing that this sample of genes possesses a distribution much different from those of other, less accessible systems affecting behaviour." (Wilson, 1975a, p.550.) And indeed, conclusions like this are things we shall have to examine later. But, even at this point it is clear that gross genetic differences are not being claimed — that is, differences out of line with what seems generally conceded.

The third, and final, point is that what is frequently overlooked is the extent to which, for all the talk of differences, the sociobiologists are affirming the unity of humankind. Take, for example, Trivers' (1971) discussion of reciprocal altruism. He draws his data about human behaviour from all kinds of society, from the least to the most technologically advanced. His claim simply is that all humans are united in having genes which lead to reciprocal altruistic behaviour. The bushman in the Kalahari desert and the New York business executive respond to other people in fundamentally the same way for fundamentally the same genetic reasons. Here, as elsewhere, a strong sociobiological theme is that for all of our Western sophistication, in many vital respects we are still sisters (or brothers) under the skin with the rest of humanity.

So much, therefore, for the direct charge that sociobiology is, intentionally or unintentionally, racist, or in some other way grossly politically reactionary. I would suggest that it is not. However, there are still some related actual or possible criticisms which must be considered. It might be agreed that sociobiology is not really racist, but it might nevertheless be felt that there is something politically pretty suspect about it. Whilst it may not endorse extreme views, it still does give spurious countenance to some Neanderlithic social and political ideas. In particular, it is just extreme Social Darwinism in modern dress. Sociobiologists may not openly quote John D. Rockefeller to the effect that the struggle for existence sanctions modern capitalism, but essentially their message is the same. Sociobiologists may not endorse racism, but they do endorse an extremely right-wing free-enterprise system.

5.2. DOES SOCIOBIOLOGY SUPPORT VIRULENT CAPITALISM?

A charge like this is to be found in Sahlins' (1976) critique of sociobiology (as well as the Boston critics' attack). Basically Sahlins sees the history of evolutionary theory, culminating in sociobiology, as having been an ever-growing circular shuttle back and forth between political economy and biology. Consequently, the ideology of Western capitalism has been infused into evolutionary speculations, which are then in turn used to provide spurious justification for socioeconomic beliefs and practices.

Thus, at one point, we have Malthus' speculations about the futility of attempting to relieve the lot of the poor:[1] given the inevitable squeeze which comes through human numbers potentially increasing at a geometrical rate whereas food supplies can be increased only at an arithmetical rate, Malthus thought that aid to the poor only increases their numbers, thus making for a yet-greater problem in the next generation. These ideas were picked up by Darwin who dropped Parson Malthus' pious mouthings about avoiding population troubles through 'prudential restraint' (*i.e.* abstinence from sexual intercourse), and who argued that in the animal world inevitably we get an all-out 'struggle for existence': this in turn providing the motive force behind natural selection.

Next, we go back to political economy as American Social Darwinians like William Graham Sumner seized on Darwin's ideas, applied them to the human sphere, and argued that they justify an extreme *laissez-faire* capitalist system. Then these ideas, permeated with the notion that the only thing that counts is profit at any cost (obviously obtained at the expense of one's own species), get translated back into animal sociobiology, with its emphasis on spreading one's genes, at the expense of one's fellow species' members, no matter how — no place here for concern for others. And so, finally, we come to the final shuttle, as these ideas are read back into the human realm and are taken to justify a thoroughly reactionary market system. Self-interest, and only self-interest, is all that counts.

But the whole process, and particularly the end result, argues Sahlins, is circular in the extreme. "Since the seventeenth century we seem to have been caught up in this vicious cycle, alternately applying the model of capitalist society to the animal kingdom, then reapplying this bourgeoisified animal kingdom to the interpretation of human society." (Sahlins, 1976, p.101.) We have read into human sociobiology precisely those ideological elements we want to read out. "Hence the response by men of the Left becomes intelligible, as does the interest of the public at large. What is inscribed in the theory of sociobiology is the entrenched ideology of Western society: the assurance of its naturalness, and the claim of its inevitability." (Sahlins, 1976, p.101.)

A number of points can be made against this criticism; but, first let us note that as history (as Sahlins himself seems aware) the objection leaves much to be desired. For instance, as was pointed out in the last chapter, the relationship between biological Darwinism and Social Darwinism is far from a simple one of historical cause and effect. Admittedly, there are times when Darwin does sound like a Social Darwinian, writing of the Caucasians beating the Turks hollow in the struggle for existence. At other times, however, he was careful to eschew any social, political, or economic extrapolations from his biological work. Moreover, there is good reason to believe that the major influence on Social Darwinism was Herbert Spencer, whose major evolutionary debt lay, not to Malthus, but to the French biologist Lamarck. (Himmelfarb, 1968.)

But let us leave this all on one side and grant Sahlins his history, which is certainly no worse than that of most scientists. The important question is: Does sociobiology incorporate and give spurious justification to Western social-political-economic ideology? Perhaps even more importantly, does acceptance of sociobiology — particularly human sociobiology — mean that we have to support and advocate Western ideology? I offer the following three points in reply to these questions.

First, let us take up the question of the flow of ideas from the human sphere to the biological world, and *vice-versa*. Now it cannot be denied that Darwinian evolutionary theory was in important respects modelled on the ideas of Western political economy. (de Beer, 1963; Ruse, 1973b.) Although Darwin knew all about the struggle for existence (even by that name) from his readings of Charles Lyell's *Principles of Geology*, his reading of Malthus' *Essay on a Principle of Population* in late September, 1838, provided the catalyst which enabled him to see that mechanism analogous to breeders' artificial selection exists universally in the wild and brings about wholesale evolutionary change. (Ruse, 1975c.) Moreover, Darwin explicitly borrowed from Malthus' ideas in his *Origin of Species*, modelling his arguments for the animal world on Malthus' related arguments for the human world. And the same is true of natural selection's co-discoverer, Alfred Russel Wallace. Furthermore, it cannot be denied that, from Darwin on, evolutionists have wanted to apply the biological theory of evolution to the evolution of humans (although curiously, primarily as a function of his enthusiasm for spiritualism, Wallace came to doubt the adequacy of natural selection as a causal mechanism for human evolution).

But, granting these facts, what in them is objectionable? What is spurious or, in some sense, a bogus justification of Western ideology? On the one hand, considering the flow of ideas into biology, as a matter of general principle the simple fact that Darwin modelled his ideas on the ideas of someone else is hardly cause for complaint. Admittedly, as a number of philosophers have pointed out, the use of models is fraught with danger — one might illegitimately read into the new area notions applicable only to the old area — and, thus, as R. B. Braithwaite cautions: "The use of models must be marked with eternal vigilance." (Braithwaite, 1953, p.93. See also Bunge, 1968.) But the use of models, at least as heuristic guides, seems indispensible in science — hardly any great discoveries would have been made without them. Moreover, in setting up a new theory modelled on something else, one is not necessarily endorsing the old theory. The whole point about a model is that in essential respects it is *not* the same as the thing on which it is being modelled. To take a simple example: Kerkulé modelled the benzine ring on his vision of a snake swallowing its tale. That a herpetologist should object that snakes never swallow their tales is irrelevant. Furthermore, still at a general level, if it be objected that Darwin's fault be not the use of models as such, but his use of models drawn from the human world, the existence of literally scores of such modellings throughout science shows at once the absurdity of such a worry. Where, for heaven's sake, did the notion of 'force' come from — or of 'work', or of 'energy', or of 'attraction', or of 'repulsion'?

Presumably at this point, the reply would be that Darwin erred, not in drawing upon the human world, but in drawing upon such a theory as that of Malthus, which advocates a theory of welfare somewhat to the right of Louis the Fourteenth. But whilst it is, indeed, true that Malthus draws a grim picture of human existence, arguing to the futility of state support for the poor, there is a fundamental difference between his picture and Darwin's. Apart from anything else, Malthus thought that his view points to the impossibility of any real change, including evolutionary change, whereas obviously Darwin did not think this about his! (Bowler, 1976.) More specifically, Darwin stripped Malthus' ideas of normative content: Darwin was not, in any sense, concerned about the rights and wrongs of the struggle for existence in the biological world, or about what we ought to do about it.

For Darwin, the undeniable universality of the struggle for existence was what mattered. And, on this score, it is hard to fault him, or for his use of Malthus. What counts in a new theory is whether it works or is right (or whatever it is that one uses to characterize successful theories), not its origins. I take it that at this point Sahlins is not questioning the essential truth of Darwinian evolutionary biology: for all that neo-vitalists like Arthur Koestler wish otherwise, the battles with saltationists and Lamarckians and others are over. Moreover, as a parting shot, I would point out that Malthus, in his turn, modelled his ideas on those of someone else, namely Benjamin Franklin's speculations about the geometrical—arithmetical conflict in the animal world! In other words, if we dig down, Darwin's roots take us away from humans.

Going at matters from the other side, considering the flow of ideas from biology back to humans, it is surely not cause for complaint — at least not cause for complaint today — that Darwinian evolutionists have wanted to apply their ideas to humans. Leaving on one side the specific programme of the sociobiologists, I take it that no one wants to deny that human beings are in some sense animals, that we are descended from monkeys (not species extant today), that natural selection has been the prime causal factor, that many of the things which distinguish us from other brutes, such as our intelligence, are adaptive and were caused by selection, and even that many of the differences between human races are not just chance, although what these exact causes might have been or whether they are still operative is quite another matter. It may perhaps even be that selection is working on humans today. Certainly, evolution is still going on, if only because of the new mutations that are always coming into the human gene-pool.[2]

Now, considering these sorts of claims in the light of Sahlins' criticisms, two comments spring to mind. First, tainted or not, they seem to be essentially true, and to be accepted by all — those for and against sociobiology. Like the Bishop of Worcester's wife, one may not much like them, but there we are. Until someone comes up with a rival theory of human evolution, one which can explain the facts as well as these do, they must suffice. But, secondly, I would deny that they are tainted with the ideology of Western capitalism. Claims like these are extensions from non-human Darwinian evolutionary biology, but, as I have just pointed out, this kind of Darwinism is shorn of the kinds of normative claims that someone like Malthus advocated. Nor have such normative claims been smuggled back again. Presumably, human ancestors with larger brains or more erect stance were selected over those with smaller brains or more crouched. This is not to say that it was better that this was so, or that we today ought to select for even larger brains or straighter backbones. Nor, for example, is it to say that if someone today goes down with a genetically caused ailment we ought to let them perish, without trying to help. And it certainly is not to give any sanction to *laissez-faire* economic beliefs and practices. To find such sanction is to read something into the evolutionary claims that is just not there. What one is doing in such a reading is setting up a model which, as Darwin's theory was to Malthus', is different from the original. In other words, at least at this level, Sahlins' shuttle view of history is without critical force.[3]

But, having dealt now with the passage of ideas from the human world to biology and back again, and found that, at a fairly basic level, Sahlins' fears are groundless, we come to the second point I want to make. Everything that I have said so far might be allowed, but still it might be felt that with the arrival of sociobiology a whole new dimension is added. Sociobiology, it might be felt, has been a fresh infusion of Western ideology, and so this, particularly as applied to humans, is suspect. Perhaps one can indeed make non-normative claims about human evolution. The point is that sociobiologists do not.

Of course, this criticism depends on the premise that such an infusion of ideology has occurred, and this is an assumption which I would question and deny. However, more of this later. Let us for the moment allow that sociobiology has given a rather different direction to evolutionary thought, whether ideological or not. Still Sahlins' criticisms fail, certainly inasmuch as they purport to show that acceptance of human sociobiology means that we must endorse right-wing Western ideology; although it must be admitted that there are times when it is necessary to save sociobiology from the sociobiologists!

5.3. WHY SAHLINS' CRITICISMS ABOUT IDEOLOGY FAIL

First, even if we agree that sociobiology is infected with Western ideology in a way that the rest of evolutionary thought is not – that it puts an added premium on self-interest, even against conspecifics – it is still legitimate and proper to make the distinction between 'is' and 'ought'.[4] Perhaps we have evolved in the way that the sociobiologists claim, but this is not to say either that we are total slaves of our genes or that we ought to sit back passively and accept that what is, is therefore right. In fact, there are good reasons for not accepting either side of this disjunction.

As we have seen, with respect to the first disjunct, no one seems to want to deny that in significant respects humans escape their genes through their culture – that is, there seems to be a dimension to human life which is not under direct control of the genes and where it is reasonable to suppose that actions might be taken which the genes unaided would not cause us to take. Perhaps stating matters a little more precisely, since I have been at pains to emphasize that in an important respect the genes are a causal factor in all of human experience, because culture is also a very important aspect to human experience we have a great flexibility in our actions: a flexibility which we would not have without our culture. Indeed, so great is the flexibility of our culture that we might take actions which could go counter to our direct reproductive interests. To mention a simple example, our genes might drive us towards maximizing our own individual reproduction, but this is not to deny that through our culture we might decide to limit reproduction for the good of the group.

As far as the second disjunct is concerned, it does not follow at all that what exists through evolution must be passively accepted as what is best or right. I shall be returning to this point in some detail in the final chapter, but its essential truth can be grasped easily. It is indeed true that that what is, or is 'natural', has a *prima facie* claim to being good. For instance, generally speaking, most of us would try to prevent a suicide because we think that life is a good thing. Similarly, to take a slightly more involved example, more and more of us are coming to think that one ought not to force homosexuals away from their sexual preferences, simply on the grounds that that is the way that they are. However, it is far from being the case that what is, or is natural, is in itself an absolute good. Many of us, for instance, would think that if someone has an incurable and dreadfully painful disease they ought to be allowed to end their lives — at least to drug themselves to a state of oblivion where they are *de facto* dead. Or, to take a less controversial example, it is 'natural' for the smallpox virus to infect humans, but no one would deny that the unnatural extinction which now faces the virus is a very good thing indeed.

In short, we do have some non-genetic power over our destiny, and what has evolved is not necessarily a good in itself. Hence, even if the sociobiology of humans be true, and even if it incorporates aspects of Western ideology, it does not follow that one is thereby committed to endorsing Western ideology as a morally worthwhile guide to future action. That conclusion requires additional assumptions: either that we have no control over our genetic fate or that what is, is therefore good.

However, I must confess that, as admitted earlier, on occasion sociobiologists (although not sociobiology!) do make these assumptions. Thus, for example, at one point, Wilson argues for "an evolutionary approach to ethics", claiming that sociobiology shows that "no single set of moral standards can be applied to all human populations, let alone all sex-age classes within each population". (Wilson, 1975a, p.564.) But, simply speaking, Wilson is wrong: sociobiology shows nothing of the sort. The fact that different people have different sex drives does not imply that different moral codes apply to them. If we found that certain genes turned men into rapists, we would (and should) certainly not sit back passively and let them go ahead. And, elsewhere, Wilson himself seems to realize this, at least by implication, for he recognizes that one of the greatest problems is the explosion of human population numbers and he urges us to do something about it — here, obviously, Wilson is not basing his morals on sociobiology. (Wilson, 1975b.) I shall return in some detail to this and related points in the final chapter of this book. (Incidentally, if we were to seek systematic bias on the part of sociobiologists, a case might be made for accusing them of anti-Catholicism, for they have no time for opponents of population growth limitations. But perhaps left-wing critics would not consider anti-Catholicism to be a bias.)

We come now to the third, and final, point I want to make against Sahlins' argument. Even if what I have thus far argued be granted, it may still be felt that politically speaking there is something very dangerous about sociobiology. Consider an analogous example: Suppose it be discovered as a matter of fact that women are physically weaker than men and that nothing we can do can alter this difference. The difference does not in itself imply a normative claim, such as "women ought not be allowed to take combat roles in the armed forces". On the other hand, it clearly does set limitations on what women can and cannot do, they may not for example be able to do all of the physical things that we call upon male soldiers to do, and therefore, if nothing else, it is liable to influence our normative positions. Why urge the creation of female combat troops, if for the same amount of time and money we can get more efficient male combat troops? Analogously, switching back to sociobiology, although a social biology of humans impregnated with right-wing Western ideology may not, as a matter of logic, imply that we ought to adopt such an ideology, its success is certainly going to turn a lot of people towards such an ideology. It would be naïve to think otherwise.

Given what I think is the obvious force to this counter-objection, it is clearly important to my stand to be able to show that sociobiology does not, in fact, paint a picture of humans as selfish, aggressive individuals, who function best as a group only when their personal desires are allowed total freedom in an open market-place, unhampered by any overriding controls such as those of the state. I believe that, in fact, one can show that human sociobiology does not paint such a picture; although as before there are times when it is necessary to save sociobiology from the sociobiologists! But more of this anon. Let us first make the case for sociobiology.

Sociobiology, even as applied to humans, is a theory essentially about the causal effects of the genes. What is claimed is that the genes give rise to behavioural characteristics, which characteristics will maximize the chances of the genes being represented in increased proportions in succeeding generations. Central to the theory is the notion of individual selection, which implies that only those genes are preserved in populations which cause copies of themselves to be perpetuated: not necessarily copies of any of the genes of the individual's population's gene-pool. In this sense, the genes are, as Dawkins (1976) graphically puts it, 'selfish'. But, of course, this is a metaphor. Genes are not really selfish, even though it may be convenient or arresting to speak this way. It is human beings who are selfish or not (and perhaps some of the higher animals). It is certainly allowed by sociobiology that the genes may give rise to selfishness, of hypocrisy, or other unpleasant behavioural traits; but then again, as we have seen, the genes may well give rise to altruistic behaviour. Furthermore, it is surely a conceptual mistake to say that because human altruism is a function of the genes (say brought about through kin selection) it is any less genuine as altruism: that there is no 'real' altruism and that sociobiology paints a picture of humans entirely selfish, as endorsed by right-wing Western ideology. If I am sincerely trying to help somebody, without any conscious thought of return — and the sociobiologists quite explicitly allow the possibility of this — then I am being genuinely altruistic. Certainly, I am being altruistic in a way not allowed by an extreme cynical view of humans, such as an extreme *laissez-faire* ideology, which sees all humans being motivated by a conscious self-interest, and which endorses this view. Because one identifies unconscious causes behind human actions, it does not follow that the actions are any less genuine or any less worthy of praise (or condemnation).

Perhaps the point I am trying to make can be brought out more clearly by noting that few today want to deny that there are some causes, conscious or unconscious, behind all human actions. This holds whether one believes that the genes have total or minimal control over human actions. Consider: all of us at some points in our lives have been punished when we have transgressed the moral norms. What was the point of that punishment? Presumably, the answer to this question, for all except the extreme retributivist, is that such punishment is intended to set up conditioning in us in such a way that we and others will not similarly transgress in the future. I will not hit my little sister again because I fear the consequences, and this gets generalized into my behaviour as an adult. In other words, genes or not, there are causes behind our actions. But this does not mean that we should throw out all evaluative terms. Certainly it does not mean that because our actions have causes we cannot distinguish between someone who is selfish and someone who is selfless. It is Pickwickian to say that we do everything out of self-interest, because then we cannot make the usual distinctions that we make. (For more on this kind of point see Hudson, 1970.)

In short, what I am arguing is that because sociobiologists have a theory of the gene in which they apply metaphors like 'selfishness', and because they think this theory explains human actions, it does not at all follow that human beings are entirely selfish or brutish or hypocritical. It certainly does not follow that humans fit an extremely reactionary Western ideological pattern. In fact, as we have seen, the sociobiologists argue that in many significant respects, humans do not fit this pattern. Although admittedly the sociobiologists allow that the genes can cause unpleasant behaviour (and who would deny the truth of this!), they also believe that because of the genes there are widespread altruistic encounters between humans: relatives and non-relatives. Moreover, it is of the essence of sociobiology that the altruism is frequently, if not usually, done without any conscious thought of return. This is in flat contradiction to the ideology that Sahlins sees the sociobiologists espousing.

One final point however: I must confess that over this whole question of the kind of picture of humanity that sociobiology paints, the sociobiologists themselves have, in respects, been their own worst enemies. Consider, for example, the following recent passage by Michael Ghiselin, which Sahlins justifiably uses to head his criticisms.

The evolution of society fits the Darwinian paradigm in its most individualistic form. Nothing in it cries out to be otherwise explained. The economy of nature is competitive from beginning to end. Understand that economy, and how it works, and the underlying reasons for social phenomena are manifest. They are the means by which one organism gains some advantage to the detriment of another. No hint of genuine charity ameliorates our vision of society, once sentimentalism has been laid aside. What passes for cooperation turns out to be a mixture of opportunism and exploitation. The impulses that lead one animal to sacrifice himself for another turn out to have their ultimate rationale in gaining advantage over a third; and acts 'for the good' of one society turn out to be performed to the detriment of the rest. Where it is in his own interest, every organism may reasonably be expected to aid his fellows. Where he has no alternative, he submits to the yoke of communal servitude. Yet given a full chance to act in his own interest, nothing but expedience will restrain him from brutalizing, from maiming, from murdering – his brother, his mate, his parent, or his child. Scratch an 'altruist', and watch a 'hypocrite' bleed. (Ghiselin 1974, p.247.)

It is undeniable that Ghiselin is making precisely the conceptual mistake on which I argue that Sahlins criticism flounders. Only in the most misleadingly metaphorical sense are altruists hypocrites under the skin. To say that someone is a 'hypocrite' is to say that, although appearing one way, they are planning to do something else. Mr. Pecksniff was a hypocrite because, whilst appearing moral, he was exploiting people like Tom Pinch. Uriah Heep was a hypocrite because, whilst professing abject humbleness, he was secretly exploiting Mr. Wickfield and scheming to trap Agnes. The whole point is that hypocrisy, like altruism, applies to individuals. Genes cannot be hypocrites, and hence inasmuch as sociobiology explains behaviour through the genes — even genes of the kinds postulated by the sociobiologists — this does not make us subcutaneous hypocrites.

All I can plead here is that sociobiology not be condemned because some sociobiologists have thought that it proves more than it does. If we are going to make that harsh a judgement, then because just about every great scientific achievement has been thought to have all kinds of fantastic non-scientific implications (as often as not by scientists themselves), we are going to end by having to jettison just about the whole of science — certainly, just about the whole of science which rises above the basic descriptive (assuming of course that there is some basic descriptive science!)

Having now completed my third and final point against Sahlins, I conclude that, on the grounds that he offers, Sahlins is wrong in finding a right-wing bias to sociobiology; although I cannot deny that some sociobiologists have thought that it proves more, ideologically speaking, than it does. There is nevertheless more ground which must be covered before we have finished with the charges that in some respects sociobiology is socially or morally offensive, and before we can turn to charges of a different kind. Even if all that has so far been argued be granted, it may still be felt that in one important respect sociobiology rather implies reactionary beliefs. This is in respect to certain oppressed sections of our society. Sociobiology, it may be claimed, rather sets up as an ideal a particular kind of human, namely a white, male heterosexual, and then, at least implicitly, unfavourably judges other humans against this model. Thus, in some sense, sociobiology undervalues blacks (and other non-whites), females, and homosexuals. In other words, sociobiology belittles many human beings in a morally offensive way.

I have little more to add about the question of racial groups than has been said already. I do not believe that sociobiology or sociobiologists imply or intend to imply that any one group of people is inferior, morally or biologically, to any other. They see differences between groups; but then, who does not? This therefore leaves the questions of women and of homosexuals (not mutually exclusive groups!) Reversing this order, I shall deal first with the rather more straightforward matter of the sociobiological explanations of human homosexuality. Then I shall turn to the somewhat more involved matter of sociobiology (and sociobiologists) and women.

5.4. SOCIOBIOLOGICAL EXPLANATIONS OF HOMOSEXUALITY

Does sociobiology put a premium on being heterosexual, belittling homosexuals? It must be admitted that a quick reading of Wilson's *Sociobiology*
might incline one to think that it does. Not only is a great deal of attention
paid to heterosexual couplings, but, more than once, homosexuality gets
grouped in with other behavioural 'abnormalities' such as cannibalism. (E.g.
Wilson, 1975a, p.255.) I am not sure that even Anita Bryant would want to
make quite this kind of evaluation! Surely here, in these kinds of musings,
we see sociobiology in its true colours? It really does have very right-wing
undertones.

Once again I must start by wishing that a little more sensitivity had been
shown by sociobiologists on such matters as these — at least wish that the
writers' true intentions had been made more obvious — for I think it is clear
that, in fact, no such implicit condemnations have been intended. Indeed, in
certain respects sociobiology and the sociobiologists paint a far from critical
picture of homosexuality.

For a start, let us note that the contexts in which Wilson lists homosexuality amongst abnormalities really do involve strange, not to say 'abnormal',
situations. Thus, for instance, he speaks of homosexuality (and cannibalism)
occurring amongst rats under conditions of very extreme crowding — incidentally, then drawing attention to similar human phenomena, such as those
which occurred in the Andersonville prison camp in the U.S. Civil War.
(Wilson, 1975a, p.255.) Now, I do not think that speaking of homosexuality
in such conditions as 'abnormal' is necessarily to make a general value claim
about the abnormality of homosexuality. What we have, presumably, are
heterosexuals engaging in behaviour which under less extreme conditions they
would eschew. This is indeed abnormal, just as it would be were one to force
homosexuals in some way to act heterosexually (or if one set up extreme
conditions which made them perform heterosexuality). 'Abnormality', in this
context, refers to behaviour which is out of the usual pattern, and makes no
moral claims at all. One wishes indeed that Wilson had made this point clear;
but it is there to be made, nevertheless.

However, this misconception now removed, we come at once to the more
important arguments: namely, those involving the actual attitudes that sociobiology takes towards homosexuality. First, let me raise and air a worry that
may be in the backs of people's minds, namely that even in bringing up the
question of homosexuality — certainly in offering 'explanations' of homosexuality — sociobiologists show a bias against homosexual behaviour. It is

rather as if sociobiologists kept referring to the size of Jewish noses and giving 'explanations' of such sizes. The simple fact of the matter is that, as problems, homosexuality and Jewish-nose sizes are pseudo-problems: they do not call for explanation and to assume otherwise reveals that one is reading into the world one's prejudices, either those based on race or those based on sexual orientation.

Now, although I realize that I will not carry some people with me at this point, I do not accept that the analogy between Jewish-nose size (which I agree is an anti-semetic pseudo-problem) and homosexual behaviour is a fair one. I personally find nothing morally offensive about homosexual behaviour as such, and I think that in an ideal world we would have such a relaxed, non-critical attitude to this kind of behaviour that it would be a matter of social indifference as to whether someone was homosexual or not: a matter of concern only inasmuch as one was relating to someone as a sexual partner. However, this does not mean that homosexuality is not a biologically interesting phenomenon, or that sociobiologists do not have a right, if not an obligation, to try to explain it. We are leaving to a later chapter full discussion of the truth of human sociobiology, but even here one can see a strong *prima facie* case for the study of human homosexuality: if evolutionary theory is indeed true and thus there is a great premium on reproductive efficiency, and if up to 10% of humans are partially or exclusively homosexual, something which seems to reduce reproductive efficiency, then there is something very puzzling here which ought to be explained. Thus I conclude that in the very act of discussing human homosexuality, sociobiologists do not show prejudice.

Turning, therefore, to the sociobiological analyses of human homosexuality, we find that that in fact there are three suggested mechanisms for such a sexual orientation. Moreover, not one of them paints an unfavourable picture of homosexuals. Indeed, inasmuch as science can properly be used to influence normative claims, the very opposite is the case.[5]

First, it is suggested that homosexuality may be a function of superior heterozygote fitness: because they reproduce less than heterosexuals homosexuals are therefore less fit biologically; but this, a result of two genes for homosexuality possessed homozygously, is kept balanced in populations by the fact that heterozygotes possessing but one homosexual-type gene reproduce more (*i.e.* are biologically fitter) than homozygotes without any homosexual type genes. (Wilson, 1975a, p.555.) In other words, the situation is postulated to be similar to that which keeps sickle-cell anaemia genes in black African populations. However, this explanation is only belittling of homosexuals if one directly equates what is biologically fit with what is morally superior. And this in itself requires a move not supplied by sociobiology: sociobiology only makes suggestions about how things are, not how they ought to be. Furthermore, it would seem that the moral inferences that socio-biologists might use such a phenomenon as balanced heterozygote homo-sexuality to support would be such as would encourage such homosexuality (at least, not discourage it). That is, given their moral positions sociobiologists would be more inclined to let such a phenomenon alone, rather than try to change it: this attitude stemming simply from the sociobiologists' fear that the population explosion is going to ruin us all. (Alexander, 1971; Wilson, 1975b.) Hence, in the eyes of sociobiology and of sociobiologists, social desirability — such as a reluctance to add to the population count — may run counter to biological fitness. Thus, in this instance, there is no condemnation of homosexuality.

The second and third suggested sociobiological mechanisms for homo-sexuality are similar. On the one hand, it is suggested that it may be a function of kin selection. (Wilson, 1975a, p.555.) Genes causing homosexuality are selected because they free the carrier from child-caring responsibilities, thus opening up the opportunities for altruistic behaviour towards kin (*i.e.* towards those who share one's genes). On the other hand, it is suggested that homo-sexuality may be a function of parental manipulation, that is that parents (perhaps unknowingly) manipulate some of their children into homosexual behaviour, in order that these children might altruistically turn to aid the reproduction of other offspring of the parents. (Trivers, 1974.) In both cases, it would seem that the actual overt behavioural practices of homosexuals involve altruistic behaviour towards others, and, whatever the causes may be, inasmuch as we praise altruism generally, homosexuals deserve praise. A whole new favourable light is cast on homosexuality. Certainly, as before, there is no condemnation of homosexuality.

Finally, let me say this about sociobiology and homosexuality. People today who condemn or otherwise abhor homosexuality tend to do so for a number of reasons. Some see it as a transgression of the moral code and a failure of free will; others see it as a sickness, which unfortunately may spread to our children (*e.g.* through homosexual teachers). As far as the first group are concerned sociobiology is at least suggesting that homosexual inclination is not a failure of moral will. Hence, it ought not be so readily condemned. After all, such people as do condemn homosexuality tend not to disparage the child with Down's Syndrome for not being very bright; perhaps they will be less ready to disparage homosexuals. As far as the second group are con-cerned, sociobiology suggests that homosexuality is not infectious. We ought therefore not fear for our children's sexual safety around homosexuals. If anything, it is heterosexuals who make people homosexual!

Either way, therefore, I suggest that sociobiology may indeed improve the lot of homosexuals, although two points should be noted. First, I must re-emphasize that I am not at this point saying that what the sociobiologists claim about homosexuality is, in fact, true. I am now assuming its truth and considering how it would fit in with our moral beliefs. Later, I shall be considering the question of evidence. Second, I am not claiming that homosexuality is an affliction like Down's Syndrome. For myself, as intimated above, whether genetic or not I consider homosexual desires rather like having blue eyes as opposed to brown. (Or more precisely, like being male rather than female. I do not particularly care that I have brown eyes rather than blue. I like being male and heterosexual; but I see no reason why someone who is neither one nor the other nor both should not perfectly well like being what they are.) What I am claiming is that those who condemn homosexuality as a failure of moral will may be less ready to do so if the causes are known. Furthermore, perhaps if they can then be persuaded to see it as less dangerous, they will not then rush in with all sorts of 'cures'.

Having now dealt with the question of sociobiology and homosexuality, let us turn at once to the question of sociobiology and women.

5.5. IS SOCIOBIOLOGY SEXIST? THE MINOR CHARGES

The Boston critics argue vigorously that sociobiology, in language and theory, unfairly demeans women: "*Sociobiology* conveys a powerful underlying statement that substantiates male supremacy in humans and other animals. The sexist message in the book is carried in subtle and implicit ways." (Allen *et al.*, 1977, p.11.) The critics argue that Wilson and other sociobiologists show their sexism because they refer to humans as 'men' or 'mankind'; they argue that the metaphors used by sociobiologists place females in an unfavourable light; they argue that higher value is placed on males, for example as in *Sociobiology* where females are described second and where females are given secondary, peripheral positions in the pictures; and, more generally, the critics complain that the whole picture that the sociobiologists draw of male—female relationships shows a bias in favour of males.

At one level, I confess that I find it hard to take some of these criticisms too seriously: certainly, they point to nothing that some routine copy-editing could not rapidly rectify. Take, for example, the use of 'man' or 'mankind'. Until recently, this was just the normal term to use to refer to human beings, and it was used happily by both males and females. Certainly, no one intended to put down women by using it, nor did women feel put down when it was used. Today, matters are in flux. Some people dislike the usage; others feel less strongly, and often dislike such ugly new compound words as 'chair-person'. But however the matter may resolve itself, it has not yet been resolved. There is no question that 'man' is as yet on a par with 'nigger' or 'yid'. In other words, at present it is difficult to read too much into the sociobiologists use of such a term as 'man', or relatedly, to the use of 'he' and 'his' as the universal pronouns. Indeed, I find that Lewontin himself was not always that sensitive, because he has not been beyond referring to evolutionists as 'he', (Lewontin, 1961, p.288), and, in 1974, just one year before *Sociobiology* was published, he quite happily referred to *Homo sapiens* as 'man'! (Lewontin, 1974, p.261.) And in any case, this criticism hardly strikes at the heart of sociobiology, for one can easily alter the language.

The complaint about the pictures in *Sociobiology* strikes me as being about as weighty as the above criticism, and about as easily answered. In fact, not all of the pictures − drawn incidentally by a woman − do put the males prominently in the centre, and even if they did this would hardly strike at the heart of the sociobiological programme or mark it as irredeemably sexist. Somewhat more serious a criticism, but thereby in my opinion when put in its true light more misleading and malicious, is the critics' charge that the sociobiologists' metaphors have 'strong sexist connotations'. (Allen *et al.*, 1977, p.11.) Thus, for example, Wilson and other sociobiologists are explicitly criticized for speaking of courtship displays as contests between 'salesmanship' and 'sales resistance'. Such a way of putting matters is considered degrading to women, as also is the claim that the courted sex will develop 'coyness'.

In order to assess the worth of this criticism, let us requote in full the pertinent passage by Wilson. He writes as follows:

Pure epigamic display can be envisioned as a contest between salesmanship and sales resistance. The sex that courts, ordinarily the male, plans to invest less reproductive effort in the offspring. What it offers to the female is chiefly evidence that it is fully normal and physiologically fit. But this warranty consists of only a brief performance, so that strong selective pressures exist for less fit individuals to present a false image. The courted sex, usually the female, will therefore find it strongly advantageous to distinguish the really fit from the pretended fit. Consequently, there will be a strong tendency for the courted sex to develop coyness. That is, its responses will be hesitant and cautious in a way that evokes still more displays and makes correct discrimination easier.

Now, as soon as matters are laid out this fully, it can be seen that the critics' charges are groundless: a veritable distortion indeed. First, it is made quite clear that it is not always the male who sells and the female who is coy. The distinction is between less or more reproductive effort. In fact, in the fish, where frequently the male raises the young, it is he who is coy and the female who does the selling. Second, it is obvious that Wilson is using metaphors. No one is really selling or buying anything – no money changes hands. What Wilson is trying to do is describe a similar situation to selling which arises in courtship. But, at least in this sense, his metaphors are appropriate. The courter has got to get its message across quickly and convincingly: this is something very much akin to selling, just as they do at car dealers' showrooms. The courted has got to look to its own interests and not get swept off its feet by an inadequate smoothie: what is this but something very much akin to coyness? Third, it is hardly that the metaphors involved, or the general description, paint that superior a picture of the male, even if we assume that it is the male sex which usually puts less effort into raising the offspring. The female does all of the hard work. The male plans on doing little or nothing, and is hiding his deficiencies into the bargain! It is surely significant, given the contempt that we tenured academics tend to feel towards those who work for commission, that Wilson should have used the term 'salesmanship', and it is surely paradoxical that, in their desire to pillory him, Wilson's critics should so far have mistaken him to conclude that he intended praise by using such a term.

At one level therefore, admittedly a fairly surface level, I do not find much in sociobiology to justify the charge of 'sexist'; although, before going on, I must confess that at this level (as seems to be so often the case) the sociobiologists themselves have not always been quite as careful as they might have been. Thus, for instance, even though not picked up by the critics, there is at least one time when Wilson himself steers dangerously close to belittling women. In a popular article, Wilson (1975b) has suggested that whereas men may have greater mathematical abilities, women may have greater verbal abilities. Now, obviously there is nothing particularly sexist in a claim such as this, although of course its truth is another matter. But then Wilson goes on to argue that this difference between the sexes may lead to men taking a stronger social role than women. Of course, unless one then assumes that being dominant is better, this again is not really sexist; but given the fact that Wilson himself offers absolutely no evidence of why his premises should lead to his conclusion, and given also the fact that his premises so obviously do not imply his conclusion, one starts to get the feeling that Wilson personally wants to derive the conclusion. Could it be that Wilson wants to imply that male dominance is natural, not the result of some conscious Machiavellian

scheming by men, and that therefore there is a *prima facie* case for assuming
that it is a good? Perhaps I am reading too much into Wilson at this point:
sensitized by the critics' charges I see sexism even where it does not exist.
Certainly, I shall be giving evidence to show that Wilson does not always
regard males highly, and obviously Wilson's lapse, if that it be, does not irrevocably flaw sociobiology as such. Sociobiology, in this instance, is certainly
not implying that women are inferior. However, I do confess that, in this
particular instance, I do not find Wilson quite as mindful of the status of
women as he might be.

5.6. IS SOCIOBIOLOGY SEXIST? THE MAJOR CHARGE

By now, sterner critics may be able to contain themselves no longer. Deliberately so far, I have been confining this discussion of sexism to a fairly surface
level, and this they may complain is precisely to miss or to avoid the point.
Forget the language, forget Wilson's possible lapses, these just scratch the
surface of the matter. The real point is that through and through, at the
deepest conceptual level, sociobiology demeans females in general and women
in particular. Thus, think back to Trivers and Dawkins on sex. The whole
analysis of sexual relations starts with the initial premise that females begin
with a disadvantage, namely that they are literally left holding the baby! And
the analysis continues with this theme of women as underdogs, as they work
to get around their initial disadvantages, either by trapping the male ('domestic-bliss strategy') or by selling themselves to the highest bidder ('he-man
strategy'). Through and through, females (women in particular) are presented
as second best, always trying to catch up with or overcome their disadvantages
as compared with males. Moreover, in the human case this leads to all kinds
of sexist stereotyping: women are the domestic animals, the child-rearers;
men are the aggressive humans, dominant over females, by nature polygamous
(polygeny), and so on. In short, the sociobiology of sexual relations is male
chauvinistic through and through.

Even against this more serious charge, I do not think that sociobiology
stands indicted of sexism, although one thing must be allowed and made very
clear. Fundamental to the sociobiology of sexual relations is the belief that
males and females are antagonists, in the sense that they are pursuing different sexual strategies. (It is only this sense of antagonism that is implied; as
will become clear, at the behavioural level males and females might cooperate
intimately and harmoniously.) Biologically speaking, males and females are
different and their ends are not necessarily the same. It is no use trying to

avoid the charge of sexism by claiming that sociobiologists do not really want to separate males and females, because the whole point of the sociobiology of sex is that this is precisely what they do want to do. Moreover, although in the case of fish it is the males who are (if anything) at a disadvantage, for humans it is females who have the child-rearing, the dominated status (if that it be), and the like. This much must be granted the critics.

But is all of this, that sociobiology would claim, necessarily sexist? I do not see that it is. Drawing attention to differences between males and females is not in itself sexist, nor even is the taking of certain actions on the basis·of these differences. I take it, for instance, that in Olympic swimming and athletics it is not considered sexist that men and women compete separately. Indeed, it would probably be considered sexist if they were forced to compete together. Men and women are different. Generally, men are bigger and stronger than women, and that is a fact of life. Sexism is using the differences between men and women to draw unwarranted normative inferences, say that because they are bigger men are better; it includes, of course, reading in fictitious differences in order to draw such normative inferences.

At this point, the critics will no doubt accuse me of quibbling or stalling. No one will deny that there are biological differences between males and females. No one will even deny that, socially speaking, males are dominant to females. ". . . it must be recognized that sexism, the *socially prescribed* power men have over women, does exist in most societies." (Allen *et al.*, 1977, p.12.) What is objectionable is the spurious authenticity sociobiologists give to this power by their suggestions of all the genetically caused behavioural differences between men and women: differences which, as we have seen, put men in a position far favourable to that of women.

As always, leaving until later questions to do with the truth or falsity of sociobiology, let me offer three reasons why, even in the face of this strong criticism, I do not see sociobiology doomed as irrevocably sexist: here, understanding sexism, as above, to imply the drawing of unwarranted normative inferences from the sexual differences between men and women, and recognizing also that until the question of evidence is fully explored we cannot give a final answer to the sexism charge. I make this latter reservation because what I shall have to say here will be predicated on the assumption that what the sociobiologists have to say about sex has fair claim to being sound and is not just prejudice disguised as objective science.

First, as pointed out already in this chapter, sociobiology is essentially a causal theory about the genes, and many of the terms used are metaphors. Frankly, there are times when I wish that the sociobiologists would ease up on their use of metaphor, or at least use less flowery ones, but metaphors should be recognized for what they are. They are not literal reality. Thus, looking at the theory, what we find is that sociobiologists argue that because some genes are passed on via sperm and some via ova, adaptively speaking they must give rise to different behavioural characteristics — the same behaviour and associated morphological characteristics will not do for both sexes. But these claims in themselves are not normative at all, and if one were determined to use them to back up normative claims they certainly do not point to the inferiority of women. The whole point about sexuality is that both sexes are absolutely essential — one sex on its own cannot reproduce. Hence, in this sense, if anything, sociobiology points to the equal worth of males and females. Moreover, as just pointed out, the different strategies the genes take are described metaphorically: there is, for instance, no question of female genes consciously demeaning and degrading themselves before male genes. Female genes do not choose he-man individuals or he-man genes. Literally speaking, what happens is that female genes give rise to characteristics that will best ensure their being linked to good male genes in the next generation (by 'good' is meant giving rise to characteristics that will most further the reproductive chances of the female genes). In making claims like this, one is not implicitly implying that males are better than females.

The second point refers to the various characteristics to which the genes give rise, since obviously the critic will at once complain that although the genes supposed behaviour may not be sexist, the physical behaviour, to which the genes are supposed to give rise, will be. First, note that by no means all of the characteristics are those traditionally associated with male chauvinism. In fact, human males (because of the domestic-bliss strategy) show considerable faithfulness towards their spouses and invest a great deal of time and energy in the raising of their offspring. Because the offspring require so much help, human males show substantial domestic virtues. This, in itself, is not normative, but it certainly does not point to a biologically determined behaviourally selfish picture of human male activity. Nor is the implication that female humans are scheming hussies, consciously planning how they can trap poor males into helping to raise their brats. The domestic-bliss strategy most definitely does not rule out the possibility that a girl or young woman might deliberately refuse at first to let a young man sleep with her for the most sincere of moral or religious reasons, permitting intercourse only after the man has shown his 'honourable' intentions through the formal act of marriage. Actually, this is just the kind of behaviour which fits the strategy.

Second, let us grasp the nettle firmly and openly admit that, for all of their domestic virtues, it is certainly implied by today's sociobiology that because of their biology men will tend to be more dominant, more polygamous, and if any avoidance of child-rearing is to occur it will be by them (conversely, women will be more coy, and so on). However, as I keep saying throughout the chapter, such claims as these are not normative in themselves; moreover, I am not even sure that they point in a direction particularly creditable to males. At least, I am not sure that being strong and he-mannish is an ideal that I particularly venerate. It all seems a bit adolescent. The kinds of human attributes that I cherish are intelligence, sensitivity, loyalty, artistic ability, and so forth. I see nothing in sociobiology which implies that women will be any less lacking than males in these respects: indeed, if anything, I should think women will be ahead. Moreover, even when it comes to something like dominance, I am not sure that the sociobiological implications are that males will be dominant in every respect. They may be dominant socially, but in their intersexual relations things might be otherwise: socially, it was Bishop Proudie who wore the apron, but, personally, it was Mrs. Proudie who wore the trousers (or gaiters!)

In fact, it is perhaps worth noting, given the charge of sexism, that for all that sociobiologists like Wilson are prepared to suggest that males (including human males) tend to be dominant in some sense, considering males generally the sociobiologists do not seem overly enamoured with them. Granting that males are, indeed, usually the initiators in courtship, we have seen how Wilson refers to their tactics as "salesmanship"; furthermore, Wilson catalogues one species after another in which, when hard work is to be done, males are conspicuous only by their absence. Significantly, of one extreme example, the lions, he refers to the males as 'parasites'. (Wilson, 1975a, p.504.) No prizing of male attributes here. Of course, one might argue that I only argue as I do because I am soaked in the Protestant work ethic. I suggest, however, that critics can argue as they do only because they have an ideal of human that combines the sex habits of Brigham Young, the macho image of Ernest Hemingway, and the work habits of Oblomov.

My third and final point to the severe charge of sexism is simply that because one argues that males and females have different biological attributes, this does not mean that one ought to sit back and quietly accept them and their consequences. Obviously, today, there are terrific cultural and technological changes affecting women. In the West, thanks to the Industrial Revolution opportunities and facilities are available to educate women, up to and including the college level. And thanks to biomedical advances, particularly in the area of contraception, women have been freed from their previous constantly child-bearing roles. Hence, whatever the biology of women (and of men), it does not follow that sociobiology implies that people ought to stay in the positions for which they were selected. Even if there were a time when women were suited for staying at home and being wives and mothers, given the changes in the environment (considered in the broadest sense) there is no necessity that the future be like the past. We ought to let human beings, males and females, develop in the fullest way possible, and thanks to modern technology, we are no longer the absolute slaves of our biology.

Of course, sociobiology does suggest a final note of caution. If men and women have different behavioural genes, one ought to be wary of assuming that men and women can happily assume exactly the same positions in this or any new society (even though one does not believe that they must be trapped in the roles of the past and one will readily concede that, as in the physical dimension, there will be considerable overlap between the sexes). Possibly, radical feminists, denying any differences and determined to create a society of total androgynous sameness, will feel that this admission really does show the sexist sting in the sociobiological tail. I do not think that it does, any more than the admission that different people have different abilities implies racism. It seems to me at least plausible to suggest that women as a whole would make better doctors than men. But perhaps we have here reached a point where further argument is fruitless. For myself, I always find it curious that radical thinkers are so conservative with respect to technology, say abhoring the use of artificial fertilizers and of nuclear power, and yet so ready to argue that we can recreate our social fabric to any form we please without deleterious side-effects. Of course, the same paradox occurs in thinkers at the other extreme. As they vote against the Equal Rights Amendment (E.R.A.), they stuff themselves with food containing all kinds of polluting additives.

NOTES TO CHAPTER 5

[1] Sahlins, in fact, refers back to Hobbes; but as he seems aware, Malthus is the launching point for biology.
[2] I shall return to, and elaborate upon, this point in the final chapter.
[3] It is of course true that, as we have seen, Darwin himself had a tendency to attach to his theory normative ideas commonly held in Victorian Britain: for instance, about sexual differences and about the superiority of white Anglo-Saxons. I am not denying that one can load down Darwinian evolutionary theory with normative claims, or that people have done so. My point, one which I think is denied by Sahlins, is that one can be a Darwinian evolutionist without being a right-wing reactionary, because the evolutionary

theory itself does not make normative claims – it does not imply the superiority of the British male.

4 Recently a number of philosophers have taken to arguing that moral claims can be derived from non-moral claims. In the final chapter, I shall be considering in some depth the extent to which scientific claims can be used for the bases of ethical claims. The discussion at this point, therefore, should be considered as somewhat preliminary, with possible objections shelved until later.

5 When I speak of science 'influencing' moral claims, I am not going back on my earlier position and now arguing that scientific claims lead directly to moral claims. Rather, I mean that scientific beliefs can influence how one thinks moral beliefs can be implemented. I might justify my moral belief that one ought to do action A rather than B, because science shows that A leads to more happiness than B. Yet, of course, the moral force of my belief does not come from the science, but from the overriding moral belief that one ought to maximize happiness.

CHAPTER 6

EPISTEMOLOGICAL CRITICISMS

So far, we have been considering criticism of sociobiology purportedly show-ing that there are things morally offensive about it. We must turn now to criticisms of a rather different kind: criticisms which purportedly show that sociobiology is a pseudo-science because it has various conceptual or method-ological defects, either internal to the subject itself or as it relates to external reality. As before, when once a criticism has been introduced, I shall feel free to comment myself.

6.1. THE PROBLEM OF REIFICATION

The first sin of which the sociobiologists are accused is what Lewontin terms as unwarranted "reification". In particular, the sociobiologists mistake the relationship between the units of heredity, the genes, and the way or ways in which these units become physically manifest through outward characteristics, organisms' phenotypes. It is argued that sociobiology is predicated on a sup-posed isomorphism between genes and physical characteristics which is crude and inaccurate: nonsensical in fact.

Geneticists long ago abandoned the naïve notion that there are genes for toes, genes for ankles, genes for the lower leg, genes for the kneecap, etc. Yet the sociobiologists break the totality of human social behaviour into arbitrary units, call these elements 'organs' of behaviour, and postulate particular genes or gene complexes for each. (Allen *et al.*, 1977, p.21.)

Thus, argue the critics, when sociobiologists try to explain things like altruism in terms of the genes, from a true genetical standpoint the whole endeavour really does not make a great deal of sense. For instance, when Trivers tries to explain altruism between non-relatives in terms of his model of reciprocal altruism promoting genes for such altruism, the whole enter-prise is essentially worthless because altruism is not a trait of a kind which can be put in correspondence with genes for that trait.

There are two points which I would make in reply to this objection, one point more general and the other more particular. First, whilst it is indeed true that organisms are not neatly divisible into an 'objective' set of traits, as

102

certain recent taxonomists have rather supposed, or that any set of characteristics is going to correspond neatly to a set of controlling genes, one must
take care that one does not get too carried away, else one will find oneself on
the path to the rejection of the whole of genetics! Let us state the obvious,
the absolutely essential precondition for the science of genetics. One can
certainly abstract characteristics from organisms, whether one believes these
to be truly objective (whatever that might mean) or not: blue eyes, curly hair,
black skin, and so on. Moreover, at least some of these characteristics can be
linked fairly directly with genes − eye colour for instance − and not only do
they follow basic laws of genetics, but physiological links between genes and
characteristics can be given. Hence, without in any sense pretending that all
abstractable characteristics can be related directly to the genes (Lewontin,
1977, instances the size of the chin, which is apparently a function of other
characteristics), it remains that some abstractable characteristics can indeed
be related directly to the genes and that these can be brought under known
laws of heredity. In itself, therefore, isolating organic units is not necessarily
a 'naïve' enterprise.

My second point brings us to behaviour: extending the abstractive process
necessary in genetics to the realm of behaviour, hoping thereby also to relate
this organic dimension to the genes, is neither naïve nor necessarily doomed
to failure. At least, this is an extension endorsed by Lewontin, for recently
he has given a long list of 'characters' or 'traits' (his language) under the
control of genes that can be selected for in Drosophila (fruit-flies), and he has
included behavioural traits! For example: "Selection for mating preference
can be carried out by allowing free mating in a mixture of two mutant stocks
and then destroying all hybrid progeny each generation." (Lewontin, 1974,
p.90.) One might add further that not only does Lewontin, in particular,
deny his own case, but that more generally the critics deny their case as it
applies to humans, for they openly concede a genetic basis for schizophrenic
behaviour, and this seems just as much to involve traits or characters as (say)
altruism.

In short, contrary to what the critics claim, there seems to be no *a priori*
incoherence involved in trying to compartmentalize behaviour into units:
units which can then be related to genes. Certainly, it might well turn out
that (say) the instances of altruism promoted by kin selection contain many
different types of altruism controlled in complicated ways by many different
genes (assuming without prejudice that altruism is genetically controlled);
but, this is a general problem for genetics, not just for sociobiology. The point
is that, as a research programme, if it makes sense to talk of fruit-fly-mating

preferences being genetically controlled, and apparently it does, it makes sense to talk of something like help given to relatives or strangers as being genetically controlled.

6.2. SOCIOBIOLOGY AS MYSTICAL NONSENSE

The second criticism comes from the pen of Sahlins and centres specifically on kin selection. In short, Sahlins considers it unscientific and 'mystical'. Obviously kin selection depends on organisms in some way sensing and, at least, acting upon fairly subtle relationships – that one can distinguish, say, a ½ genetic relatedness from a ¼ genetic relatedness (by 'distinguish' in this context is certainly not implied a conscious distinguishing). Sahlins finds this fact to make kin selection, not so much false, as incoherent. He writes:

... it needs to be remarked that the epistemological problems presented by a lack of linguistic support for calculating r, coefficients of relationship, amount to a serious defect in the theory of kin selection. Fractions are of very rare occurrence in the world's languages, appearing in Indo-European and in the archaic civilizations of the Near and Far East, but they are generally lacking among the so-called primitive peoples. Hunters and gatherers generally do not have counting systems beyond *one, two* and *three*. I refrain from comment on the even greater problem of how animals are supposed to figure out how that r [ego, first cousins] = 1/8. The failure of sociobiologists to address this problem introduces a considerable mysticism in their theory. (Sahlins, 1976, pp.44–5.)

He then goes on somewhat to 'soften' his criticism by suggesting that at least kin selection runs close to the fallacy of affirming the consequent, namely (Sahlins quoting Wilson!) "constructing a particular model from a set of postulates, obtaining a result, noting that approximately the predicted result does exist in nature, and concluding thereby that the postulates are true". (Sahlins, 1976, p.45 quoting Wilson, 1975a.)

This criticism by Sahlins, one which I presume would not be accepted by the Boston critics since they accept in principle, and, in limited extent, in fact, the mechanism of kin selection, shows either an intention to apply to biology far stricter standards than one would apply to the physical sciences, or more probably ignorance about the nature of scientific theories and their confirmation. Take the stronger part of the criticism. It is indeed true that sociobiologists do not have much idea about how Hymenoptera mothers might sense that they are more closely related to their sisters than to their daughters, although pretty obviously it will not be through some conscious calculation of ratios as leads us to discern the different relationships. But belief that there are such mechanisms of sensing (perhaps going beyond our five senses), despite admitting ignorance about the nature of such mechanisms, is not tantamount to mysticism. This is so for at least two reasons.

First, scientists frequently do not know all of the mechanisms that their theories presupposed. Take, for instance, Mendelian genetics. When it was initially developed, no one really had much idea about how a gene could affect a characteristic or how a mutation could alter things. But there was nothing very mystical about geneticists being Mendelians: indeed it was rational to be so, because by virtue of their theory Mendelians were spurred to discover mechanisms, a research programme which has borne such wonderful fruits at our present time. Hence, that sociobiologists do not know their mechanisms is cause for cheer for problems to be solved rather than complaints of quasi-religious musings.

Second, it is hardly the case that sociobiologists have just turned their backs on the nature of these mechanisms: already they are studying in some detail the various ways and levels at which organisms can gather information about the external world, particularly about members of their own species. Wilson (1975a), for example, devotes considerable attention to the problems of communication, and certainly for the lower organisms he details the way in which much information about fellow species' members can be gleaned through pheromones, that is to say various kinds of chemical substances used for communication. It is clear that we are just on the threshold of matters here and that Sahlins' pessimism is quite premature.

In short, Sahlins' charge that the use of kin selection is in some way nonsensical strikes me as being like that of a critic who accuses an ornithologist of mysticism, because the ornithologist insists that birds are guided in their migrations even though he does not know the nature of the guiding mechanism (except that, as in the kin selection case, it is not by using human means such as looking at road maps!)

There still remains Sahlins' 'softer' criticism, namely that which centres in on the way in which kin selection might be tested. What Sahlins puts his finger on, of course, is a problem with, or at least a feature of, all scientific theories, not just sociobiology. One cannot offer logical, that is deductive, justification of scientific theories. Although the quotation from Wilson rather implies that only in bad science does one get a kind of backwards progression from fact to theory (which implies the facts), in reality this is the essence of all scientific justification. If possible, one matches one's predictions or implications to the facts, hoping thereby to add weight to one's theory, but even if there is a match, one is not absolutely guaranteed that one's theory is true — there could always be new falsifying facts or rival theories. It is for this reason that Popperian philosophers prefer to regard the mark of science as its falsifiability rather than as its confirmability, because one can never be certain of truth. (Popper, 1959.) In a sense, therefore, in trying to confirm any theory by pointing to its true predictions one is indeed committing the fallacy of affirming the consequent: in fact, it is a standing joke amongst philosophers that the first halves of logic books, the deductive halves, point out all the fallacies, and that the second halves, the inductive halves, teach one how to commit the fallacies pointed out in the first halves!

Of course, matters are not really quite this extreme. Leaving aside for a short while the question of falsifiability, blunt talk of 'fallacies' without qualification is rather misleading in this context. Science is just not open to deductive justification, even though it has taken both scientists and philosophers a long time to realize this. We have to work tentatively, working with what seems to fit, at least until something better comes along. Further, there are all sorts of considerations which lift scientific theory testing above the level of crude, fallacious reasoning. Theory testing appeals to such things as simplicity, the extent to which one can make surprising new predictions in areas thought hostile to one's theory, the extent to which one can tie together different areas of inquiry, and so forth. (Hempel, 1966; Bunge, 1967.)

Now, this is what holds for physics and chemistry, and, generally speaking, people seem to think that it is all fairly adequate. For instance, although there are many unresolved questions on the boundaries, no one seriously doubts that the earth is not the centre of the universe or that the chemical composition of water is H_2O. Certainly, I would not want to suggest that anything in sociobiology is as well established as these claims; indeed, at this point, I am not even suggesting that human sociobiology is all that plausible. But I do think it is the case that insofar as the key to the confirmation of the rest of science lies in matching predictions against the facts (supplemented by considerations of the kind mentioned just above), the same courtesy ought to be extended to sociobiology. Inasmuch as Trivers weighs ants to see if his rather surprising predictions about kin selection obtain, or Alexander searches anthropological data to see if his mothers' brother hypothesis holds up, sociobiologists are employing the methodology of all scientists. There is no call for talk of 'epistemological problems'.

6.3. NATURAL SELECTION AS SOCIAL EXPLOITATION

The third criticism aimed at showing that there is something conceptually suspect about sociobiology comes also from Sahlins. It is a kind of epistemological version of an argument we have seen already from him regarding the socio-economic implications of sociobiology. In particular, Sahlins argues that sociobiology is not, as I have suggested earlier, just an extension of orthodox evolutionary theory, something which does not break with the accepted 'paradigm'. In fact, Sahlins thinks that it does break with the past and that it gets impregnated with all kinds of Western ideological socio-economic ideas, that affect it not just normatively (as discussed earlier) but in its very metaphysical or epistemological roots. And this leads to all kinds of inadequacies and contradictions.

Thus, Sahlins writes:

The Darwinian concept of natural selection has suffered a serious ideological derailment in recent years. Elements of the economic theory of action appropriate to the competitive market have been progressively substituted for the 'opportunistic' strategy of evolution envisioned in the 1940s and 1950s by Simpson, Mayr, J. Huxley, Dobzansky, and others. It might be said that Darwinism, at first appropriated to society as 'social Darwinism', has returned to biology as a genetic capitalism. Sociobiology has especially contributed to the final stages of this theoretical development. (Sahlins, 1976, p.72.)

As a result of this, argues Sahlins, we have a new reading of natural selection, one which has "a different calculus of evolutionary advantage than did the traditional idea of 'differential reproduction' ". (*Ibid.*) In particular, organisms go after each other and after each other's goods (understood in a broad sense), rather than making do with the environment. "In the last stages of ideological derailment, sociobiology conceives the selective strategy — insofar as it is played out in social interactions — as the appropriation of other organism's life powers to one's own reproductive benefit. Natural selection is ultimately transformed from the appropriation of natural resources to the expropriation of others' resources." (Sahlins, 1976, p.73.) In short, natural selection becomes "social exploitation". (*Ibid.*)

Of course, this may all be very true and to left-wing thinkers rather sad, but what difference does it all make? If we can strip away the normative aspects, if such there be, as I have suggested we can, why should anyone be concerned? Simply, argues Sahlins, because this new theory will not work; and in support of his position Sahlins looks in detail at some of the proposed mechanisms of sociobiology, for instance Trivers' model of reciprocal altruism, and suggests that they contain internal contradictions. Inasmuch as Trivers' arguments succeed at all, argues Sahlins, it is because he reverts back to group selective ideas, away from the individual.

Once again, I would suggest that the criticisms have little or no force. Let us take first, and protest, Sahlins' entirely bogus history of biology. It is just not true that sociobiology represents a radical deviation from orthodox Darwinism. Right from the start, Darwin recognized that the struggle for existence, and the consequent natural selection, would involve not just nature, but other organisms, even — perhaps most desperately — organisms most like oneself (*i.e.* members of one's own species). "Hence, as more individuals are produced than can possibly survive, there must in every case be a struggle for existence, *either one individual with another of the same species*, or with the individuals of different species, or with the physical conditions of life. It is the doctrine of Malthus applied with manifold force to the whole animal and vegetable kingdoms . . ." (Darwin, 1859, p.63, my italics.) If killing another organism for food, or for space, or whatever, does not involve "the appropriation of other organisms' life powers to one's own reproductive benefit", I do not know what does. And incidentally, it has been plausibly suggested by a recent scholar that Darwin's emphasis on intra-specific competition is basically peculiar to him, for Malthus only really saw such competition in primitive tribes — for more civilized races the selective pressures came more from outside. namely as starvation brought about by lack of food. (Bowler, 1976.)

Moreover, emphasizing the continuity between Darwin and the sociobiologists, there is Darwin's mechanism of sexual selection, something entirely between members of the same species, and, as we have seen, an important influence on modern sociobiological thought. There is Darwin's refusal to follow Wallace in accepting group selective mechanisms in order to explain the evolution of physiological sterility, insisting that selection must act on and through the individual. And more than anything, there was Darwin's desire to apply his principles to man. (Darwin, 1871.) In other words, considering the beginning and end links, there is strong continuity between Darwin and the sociobiologists.

I do not, of course, want to pretend that everything in sociobiology can be found in Darwin.[1] Although Darwin and the sociobiologists share a fascination with the sterile castes of the Hymenoptera, whereas Darwin explained such phenomena by assuming that the family could be considered as an individual, the sociobiologists argue that mothers and daughters are biological antagonists. And it is also true that the important evolutionists of the 1930's and 40's, Mayr (1942), Simpson (1944, 1953) and Dobzhansky (1937), were not directly concerned with things like kin selection; although one must recognize what their task was, namely to put flesh on the mathematical work of Fisher, Haldane, and Sewell Wright, the men who synthesized Darwinian selection and Mendelian genetics. (Provine, 1971; Hull, 1974.) Hence, the great evolutionists who articulated and made plausible the synthetic theory of evolution, were concerned primarily with aspects of evolution more obvious and tangible than behaviour, morphological characteristics for instance. But the continuity from Darwin to sociobiology was never lost. Fisher (1930) and Haldane (1955), for instance, both grasped the theory of kin selection. And when the work of synthetic theorists touched on behaviour, one finds pre-sociobiological ideas. A classic example is David Lack's individual selection explanation of clutch-size in birds, as expounded in an earlier chapter. (Lack, 1954, 1966.) Perhaps the best evidence of the way in which sociobiology has theoretical roots into the past lies in the fact that, for all of their dislike, the more biologically sensitive critics of sociobiology usually grudgingly allow it some theoretical and factual legitimacy. Lewontin, for example, accepts the existence of at least some kin selection in the animal world (Allen *et al.*, 1977) and Levins, who has tried to prove that group selection can sometimes occur, starts with the presupposition that it can occur only under very peculiar circumstances. (Levins, 1970.)

At this point, some readers may complain about what they take to be the irrelevance of this discussion. What counts is not the history of sociobiology, but its present. I am not entirely sure about the force of this complaint. If one can, in fact, show, as I think one can, that sociobiology (including human sociobiology) is a natural and unforced growth from orthodox evolutionary theory, then obviously sociobiology can bask in some of the glory of evolutionary theory. But let us now turn to what Sahlins sees as the implications of the wrong-headed epistemology of sociobiology, noting incidentally that it was Sahlins, not I, who introduced history into the discussion in order to give his case spurious plausibility.

Let us take the mechanism of reciprocal altruism, which Sahlins singles out for particular critical fire. Sahlins writes:

The decisive point is that Trivers becomes so interested in the fact that in helping others one helps himself, he forgets that in so doing one also benefits genetic competitors as much as oneself, so that in all moves that generalize a reciprocal balance, no *differential* (let alone optimal) advantage accrues to this so-called adaptive activity. In the name of adaptation, the virtue attributed to the development of reciprocal altruism is that it eliminates differential individual advantage all along the line. (Sahlins, 1976, p.87.)

Then, perhaps somewhat condescendingly, Sahlins goes on to say:

Actually, what Trivers produces is a very good model of "group selection" or, as it might better be called, "social selection". In this model, moreover, the unit of selection is not the individual organism, nor strictly speaking is it the group, but certain *social relations* into which individuals enter in pursuing their own lives. These relations may not confer any differential advantages to these individuals taken separately. But they do advantage the group or sub-group practicing them, thus indirectly the individuals participating in them, vis-a-vis others of the species who might be incapable of entertaining the relations in question. Even for the biological study of animal social organization, it will be necessary to take a "superorganic" perspective. Meanwhile, as for the biology of reciprocal altruism, the perspective of sociobiology collapses under the contradiction that such generalized altruism yields no *differential* benefits in individual fitness. (Sahlins, 1976, pp.87–88.)

In this passage, it is clear that Sahlins exhibits a fundamental misconception of evolutionary theory: that is, of evolutionary theory of the most orthodox kind. And this is quite apart from the fact that in his eagerness to criticise, Sahlins entirely omits mention of the fact that even in reciprocal altruism Trivers makes quite explicit that selection will favour any behaviour that will indeed give its possessor an advantage over its fellows, for example subtle forms of cheating. It is just not true that reciprocal altruism "eliminates differential individual advantage all along the line". More importantly, however, the fact of the matter is that, sociobiology quite apart, selection can and does act not merely to raise an individual above its fellows, but also to keep an organism up to the mark of its fellows. Differential reproduction means not merely that the superadvantaged will be favoured, but that the subadvantaged will be disfavoured. If a group of organisms practice reciprocal altruism, then it will pay an individual to be a reciprocal altruist, even though in so doing it helps its fellows, simply because not to do so will hurt the individual more.

The models, based on Maynard Smith's notion of an ESS, demonstrate this fact quite clearly, for the situation in reciprocal altruism is analogous to limited aggression: it pays an organism not to be an all-out hawk, despite the fact that its fellows suffer less, because a hawkish strategy means that the organism itself suffers more. Sahlins' criticism shows that he is quite unaware of facts such as these. Indeed, were his criticism well taken, then a fairly obvious generalization would show the impossibility of the evolution of any characteristic which did not raise an organism above its fellows – such as the eye or the hand. And this conclusion is ridiculous.

In sum, there is nothing inherently paradoxical in Trivers' model of reciprocal altruism, and there is certainly no suggestion of a covert reintroduction of group selection. One may perhaps decide to reject the model on empirical grounds, or one may perhaps accept it for animals but reject it for humans; but these are quite other matters. As a model for the evolution of behaviour, reciprocal altruism has not been shown to be in conflict with accepted evolutionary principles.

6.4. IS SOCIOBIOLOGY UNFALSIFIABLE?
GENERAL CONSIDERATIONS

We come now to the major charge against sociobiology as a genuine science: a charge levelled both by the Boston critics and by Sahlins. Genuine science, they all argue, must, in some sense, lay itself open to the test of empirical experience; that is, it must in principle be falsifiable. But it is clear that sociobiology is not, in fact, falsifiable; hence, in this all-important respect it shows itself not to be a genuine science.

Thus, for example, the Science for the People group argues as follows:

When we examine carefully the manner in which sociobiology pretends to explain all behaviours as adaptive, it becomes obvious that the theory is so constructed that *no tests are possible*. There exists no imaginable situation which cannot be explained; it is *necessarily confirmed by every observation*. The mode of explanation involves three possible levels of the operation of natural selection: 1. classical individual selection to account for obviously self-serving behaviours; 2. kin selection to account for altruistic or submissive acts toward relatives; 3. reciprocal altruism to account for altruistic behaviours directed toward unrelated persons. All that remains is to make up a 'just-so' story of adaptation with the appropriate form of selection acting. (Allen *et al.*, 1977, p.24.)

And very similar sentiments are echoed by Sahlins (1976) who, in fact, notes with approval the stand taken by the Boston group.

Once again, I believe we have a criticism which is nowhere like as devastating as might appear at first sight. However, I do not believe that the proper way to answer it is simply by taking it on its own terms: that way rather gives the critics half their case. First, a couple of preliminary points must be made: one about falsifiability in general and the other about evolutionary theory in general.

For some reasons, not entirely clear to me, 'falsifiability' seems to be a popular cry with scientists these days. The only true mark of science, they all echo, is that it could possibly throw up implications that would prove empirically false, and thus logically imply the falsity of the scientific theory itself. (See for example Ayala and Dobzhansky, 1974.)

Relatedly, Sir Karl Popper who has made so much of the concept of falsifiability in his various writings has rather become a patron saint of science, to be invoked and praised in much the way that the Great Men of Science of the Past are invoked and praised. I suppose that part of the reason for the

success of the notion – apart from the fact that Popper is one of the few
philosophers of science who have made a genuine effort to communicate with
scientists – is that it is relatively simplistic and that it makes scientists look
rather good. Unlike ordinary mortals, they do not hang on desperately to
favoured brainchilds, but (in Thomas Henry Huxley's words) sit down before
the facts as if a little child, ruthlessly rejecting when the facts say 'no'. People
of utmost rationality.

In fact, as Popper (1959) himself fully admits, matters are not quite as
simple as all this. Suppose one has a hypothesis, that it leads to false consequences, and that hence on a naïve view of falsificationism the hypothesis is
false. There are all kinds of things that scientists can and *do* do in order to
avoid throwing out their hypothesis (as Duhem, 1914, pointed out). For a
start, hypotheses are rarely tested in isolation – within any interesting scientific theory there are usually a group of associated hypotheses – and so, if
one is really keen to hold on to one particular scientific hypothesis, one can
always (or nearly always) blame an unsuccessful prediction on one of the
associated hypotheses. Or one can invoke some *ad hoc* hypothesis, perhaps
designed to show why falsification was not achieved in this particular case.
Or, if all else fails, one can always deny or forget the opposing facts. (See
Hempel, 1966.)

I am not saying that scientists ought always do one or more of these
things. I suspect that most of us would be inclined to say that someone like
Velikovsy, who does all three, rather oversteps the mark. There comes a point
when one really ought to concede that the facts decide against a theory. However, it is clear that doing some of these things does not necessarily make one
irrational or a bad scientist. For example, it is notorious that the world's
greatest theory ever, Newtonian mechanics, never worked exactly. At all
times there were always facts, for instance the perihelion of Mercury, which
never quite fitted the predictions. But it does not mean that Newtonians were
wrong or outside science when they held on to their theory despite these
facts. Nor were the Darwinians wrong to hold on to their theory, despite the
fact that for the first fifty years of its life there was the clearest possible
evidence that the time-span of the Earth was far too short for so leisurely
a mechanism as selection (the physicists underestimated the earth-span because they did not know of radio-active decay!) (Burchfield, 1975.)

A great deal of ink has been spilt on the question of when a scientist ought
properly refuse to let her or his theory be falsified, and when the time comes
when one should let the theory go. Some obvious pertinent considerations
include whether or not one has any substitute, more-satisfactory theory for

the about-to-be-rejected theory; how successful the theory has been to date in explaining difficult facts, and showing precision and in uniting different areas of research; how well the theory meshes with other theories one has; and perhaps certain metaphysical considerations, namely does the theory fit in with our views about how theories ought to be (*e.g.* does the theory conform to our accepted notions of causality?) Here is not the time or place to evaluate these various considerations, but one thing is surely clear: one cannot, or at least ought not, isolate one or two elements from a scientific theory and talk about whether or not they show a theory to be falsifiable, or rather falsifiable in a proper sense. To seize, say, on the perihelion of Mercury and its failure to fit predictions, and then to make generalizations about the falsifiability of Newtonianism is to take matters out of context to the point of distortion. One must consider the whole picture. And this obviously applies equally to the case of sociobiology.

This now brings me to my second general point, namely that concerning evolutionary theory as such. Judged in the light of some of the things that have just been said, is there something suspect about evolutionary theory, namely that notwithstanding all the other considerations that enter in, there is an essential failure by Darwinian evolutionary theory to let itself be exposed to the test of experience? Clearly, this is a question that must be answered before we can turn to sociobiology in particular. Because it is obvious that Newtonian mechanics judged as a whole was falsifiable in a proper sense, it is clear that the perihelion-of-Mercury matter did not put the whole theory beyond the realms of decent science. Similarly, we must know whether Darwinian evolutionary theory judged as a whole is falsifiable in a proper sense.

Now, it cannot be denied that many writers have argued that, in fact, Darwinian evolutionary theory is not a genuine theory because it is not properly falsifiable. Indeed Popper himself inclines to this school of thought: "I have come to the conclusion that Darwinism is not a testable scientific theory but a *metaphysical research programme* – a possible framework for testable theories." (Popper, 1974, p.134, his italics.) Usually criticisms of this kind, including Popper's, are founded on a belief that the key Darwinian mechanism, natural selection, translates into 'the survival of the fittest', and that since the fittest are by definition those that survive, natural selection cashes out as an empty tautology: it cannot be proven false by experience. In fact, however, it is not that difficult to see that this argument is specious: natural selection does, in fact, make significant empirical claims. First, it makes the claim that there will be a differential reproduction at all: this could be false for all organisms could have exactly one and only one offspring (living the same length of time and so on). Second, it claims that the differential reproduction will be *systematic*: it is not just chance which organism survives and reproduces, and there is continuity across situations. Again, this is something which could be false.

More generally, it is clear that evolutionary theory is open to the test of experience. No less an authority than Lewontin has written: "Evolution is the necessary consequence of three observations about the world ... They are: (1) There is phenotypic variation, the members of a species do not all look and act alike. (2) There is a correlation between parents and offspring ... (3) Different phenotypes leave different numbers of offspring in *remote* generations ... These are three contingent statements, all of which are true about at least some part of the biological world ... There is nothing tautological here." (Lewontin 1969, pp.41–2, his italics. See also Hull 1974; Ruse 1973, 1977d.)

But still one might be worried about the whole question of adaptation. Evolutionary theory argues that organisms are 'adapted' to their environment, that is they have 'adaptations' which help them to survive and reproduce — characteristics like the hand and the eye seem as if they were designed to help their possessors in the struggle to survive and reproduce. Surely in arguing for adaptations, evolutionists show that their theory is unfalsifiable: if the adaptive advantage can be shown then evolutionists are right, and if it cannot be shown then they assume it is there anyway! The situation is similar to when someone makes the metaphysical claim that 'God is love': if things go well then God is love, and if things go badly then we cannot yet see how indeed God is still showing his love. (Rudwick, 1964.)

A number of comments are in order here. First, it is just not true that evolutionists believe that all characteristics are adaptive. In fact, they list a number of reasons why characteristics in certain circumstances might not be adaptive. There is genetic drift, namely the possibility that slightly deleterious characteristics might get established in a population through accidents of mating. There is pleiotropism, namely the possibility of two characteristics, caused by the same genes, being selected for, because one is so advantageous that it is worth having even though the other characteristic is in some sense maladaptive. There is allometric growth, where a characteristic matures rapidly giving its possessor an early advantage in reproduction, but in later life proving positively harmful. And there are other possible reasons also. (Simpson, 1953.)

Second, linking up the discussion with some of the things that were said earlier about falsifiability, there are good scientific reasons why evolutionists are justified in assuming adaptive advantage even where they might not be able to tell what it is. On the one hand, evolutionary theory itself is too good to throw out as soon as one faces problematical cases: it unifies widely disparate areas of study, from biogeography to embryology; it leads to worthwhile predictions; it has no rivals to replace it; and it is metaphysically acceptable in that it tries to explain adaptation through normal laws rather than creative interferences by God. On the other hand, assuming adaptive advantage is a good heuristic guide — it directs evolutionists to look for the precise nature of the adaptive advantage, and has, in fact, often paid rich dividends, for such advantage has been found even though initially it seemed totally lacking.

Indeed, I would say that so useful a guide has it proven to assume adaptive advantage, that today evolutionists more readily assume adaptive advantage than they did, say, twenty years ago. Then, many seemingly meaningless characteristics were put down to drift; but now these very characteristics are thought with good reason to be paradigm examples of the effects of selection. One thinks here, for instance, of the banding on snails, which was thought valueless and is now known to have crucial camouflage significance. (Sheppard, 1975.) In short, because their theory has so often proven right or fruitful in the past, when faced with unknown or seemingly counter-evidence, today's evolutionists are right not to succumb too quickly to falsificationist blues.

Third, still answering the charge that their concept of adaptation shows that their theory cannot be falsified, I suspect in any case that examples can be created which would give evolutionists pause for thought: where they really might think that their theory would demand at least some kind of revision. Suppose one came across a group of animals with one leg grossly swollen (as with elephantiasis). Suppose that no good reason could be found for the leg, and good reasons against it could be found: the leg is not sexually attractive, it makes the bearer far more prone to be caught by predators, and so on. Suppose that there was no evidence that the leg is diminishing in size, if anything it seems to be growing (and there is clear evidence that it is genetic in origin). I would imagine something like this would worry evolutionists, especially if it seemed not to be an isolated case, but of a fairly common pattern. Certainly, evolutionists would need to rethink their premises very carefully. Of course, that such things do not often occur, does not mean that evolutionary theory is unfalsifiable. One should not confuse not being falsified with being unfalsifiable. Evolutionary theory might be true!

Let us add up the discussion so far and then turn to sociobiology. First, we have seen that the whole question of falsification is a complex one and that there might be good reasons why scientists would and should not throw up a theory as soon as some problematical evidence is unearthed. Although no one would want to deny that the ultimate test of a scientific theory is empirical experience, and that if a theory persistently fails to fit experience it must ultimately be rejected, science would get nowhere if whenever contrary evidence is obtained, scientists at once throw out their theories. Second, we have seen also that judged by these reasonable standards evolutionary theory seems a genuine theory. Moreover, it is reasonable for evolutionists to suppose that most organic characteristics have an adaptive function, or, as with pleiotropism, are in some sense linked to an adaptive function, and for them therefore to use this supposition as a heuristic guide when faced with new or undiscussed organic phenomena.[2]

6.5. IS SOCIOBIOLOGY UNFALSIFIABLE?
PARTICULAR CONSIDERATIONS

Obviously, already we can draw some important implications for sociobiology with respect to the whole question of falsifiability. Inasmuch as sociobiology is an extension of orthodox evolutionary theory, and we have seen reason to believe that it is, sociobiologists might reasonably — indeed ought — assume that the organic characteristics with which they deal, namely behavioural characteristics, have an adaptive significance. Or that such characteristics are in some way linked to adaptive functions. To make this assumption is not illegitimately to turn the sociobiologists' work into unfalsifiable pseudoscience; it is rather what they as good evolutionists ought to assume. In other words, the somewhat cavalier attitude taken by the critics is misleading. The attempt to relate behaviour to adaptive advantage is quite proper.

This now brings me to particular features of the sociobiological programme and to the critics' specific charges. None of these seem to have too much force. The main objection that the critics appear to have is that the sociobiologists rely on three different mechanisms, individual selection, kin selection, and reciprocal altruism. Now, if this is indeed an objection then it is a bit ridiculous. We are dealing with complex matters in behaviour, and it is hardly surprising that these matters are not straightforward or that more than one mechanism is needed. It would hardly do for sociobiologists to stick (say) with good old-fashioned individual selection, because they could then not start to explain kin interactions — animal or human. That sociobiology tries to provide answers to different aspects of behaviour is hardly a failing: it is a virtue. (A word of elucidation may be appropriate here. I have emphasized that today's evolutionists think that virtually all selection is for the individual, as opposed to the group. In this paragraph obviously 'individual selection' is being used in a more narrow sense, implying selection which concerns direct benefits to the individual, as opposed to indirect benefits through other individuals.)

Undoubtedly, the objection of the critics to sociobiology on grounds of unfalsifiability will be recast. It will be claimed that the real objection is not so much that sociobiology relies on various mechanisms, but that it relies on the mechanisms that it does in the way that it does. However, once again there is hardly cause for complaint, at least not on the falsifiability charge. It is certainly not the case that kin selection, for example, is some *ad hoc* invention, without foundation, invoked merely to save sociobiological face. We know that it comes from fundamental principles of evolutionary theory, and indeed its reasonableness and existence is accepted by the critics themselves! (Allen *et al.*, 1977.) Moreover, in both the animal and human world it seems to lay itself open to possible falsification. If the ratios and practices it predicts do not obtain, then it is false. Certainly, it is false inasmuch as one wants to claim that it is applicable to a wide range of interesting phenomena. (One can always save a hypothesis by denying that it was intended to apply to some particular falsifying case. However, whilst this is an allowable gambit — Boyle's law is not intended to apply to high temperatures and pressures — if one overplays it, then one is liable to end up with no real-life applications.)

Thus, for instance, as we know, Alexander has cast grave doubt on the efficacy of kin selection amongst the Hymenoptera. (Alexander and Sherman, 1977.) He suggests that the empirical evidence just does not support the claim made by Trivers and Hare that kin selection has been a major evolutionary force in determining sex ratios in many species of these social insects. This, it seems to me, is exposing a theory to a test no less than if we conclude that Snell's law is falsified after applying it to Iceland spar. Moreover, it is worth noting that the sociobiologists themselves seem open to persuasion by contrary evidence. Wilson, for instance, was initially most impressed by Trivers and Hare's results; but now, in the face of Alexander's criticisms, he has considerably tempered his enthusiasm. (Oster and Wilson, 1978.)

Much the same logical situation with respect to kin selection seems to obtain in the human world. Alexander, for instance, suggests that when paternity is in doubt we might reasonably look for the mother's brother pattern of child care.

... so long as adult brothers and sisters tend to remain in sufficient social proximity that men are capable of assisting their sister's offspring, a general society-wide lowering of confidence of paternity is predicted, on grounds of kin selection, to lead to a society-wide

prominence, or institutionalization, of mother's brother as an appropriate male dispenser
of parental benefits. (Alexander, 1977, p.17.)

Alexander claims that, in fact, his prediction does turn out to be true; but if
this is not a case of someone leading with their epistemological chin, I do not
know what is. If the facts pertaining to mother's brother care were otherwise,
or prove to be otherwise on closer scrutiny, then Alexander's position is
falsified. I should add, incidentally, that with respect to this question of
human kin selection I am not sure how seriously one should take the charge
of unfalsifiability, for, as we shall see shortly, Sahlins thinks there is good
evidence proving it false.

Exactly the same points just made about kin selection seem to hold for
reciprocal altruism. If an organism gives and gives to a non-relative, and there
is absolutely no gain and only loss, then claims about reciprocal altruism as an
effective mechanism are falsified – and this holds of the animal or human
realms. And the same is true of the relationship between the various selective
mechanisms. Alexander (1975) argues that kin selection and reciprocal altruism should and do match the anthropological reciprocity divisions marked by
Sahlins. If they do not, then we have falsifying material.

The critics' particular charges seem therefore to be without much force.
But still there might be felt a sense of unease. Somehow, it may be argued,
the sociobiologists are just a big too quick to see adaptive advantage in all
behaviour, particularly all human behaviour. Perhaps they edge close to unfalsifiability here. Already I have really answered this charge – as orthodox
evolutionists, the sociobiologists are committed to looking for adaptive advantage. One expects them to do just this. It might be pointed out however,
that, in fact, like other evolutionists, contrary to what the critics claim, the
sociobiologists do not absolutely insist that all behaviour, even all human
behaviour, even all human behaviour under the control of the genes, must
have immediate adaptive benefits. Remember, for instance, that one suggested
mechanism for human homosexual behaviour was balanced heterozygote fitness. It was openly argued here that, biologically speaking, homosexuality is
not an adaptation. And, more generally, there seems no reason to suppose
that, any more than other evolutionists, sociobiologists would refuse to allow
the possibility of any of the causes which might bring about non-adaptive
characteristics.

Coming towards the end of this discussion of falsifiability, what I would
conclude is that no sound case for the genuine unfalsifiability of sociobiology
has been established. Inasmuch as it is proper to regard falsifiability as an

essential ingredient of the scientific endeavour, sociobiology passes the test. I would add, moreover, that the case has been made by considering only one part of the sociobiological corpus, namely that centering on altruism. When the full range of its claims is considered, the scope for falsification gets broadened. Consider, for example, just the sociobiological case for human incest taboos, namely that such taboos are fashioned by selection because close inbreeding has such deleterious effects. Suppose anthropologists found that certain tribes, despite dreadful genetic consequences, openly and deliberately practiced incest for generation after generation – indeed looking upon those who did not mate with close relatives as pariahs. Suppose, also, that anthropologists found that these tribes had changed from more usual mating patterns, without any great reason (and that hence there was no question of change of genes). The sociobiological claim that incest taboos are genetic in origin not just cultural would obviously have been falsified. Here, as in the cases discussed earlier, sociobiology does not fall down as a genuine science.

6.6. IS HUMAN SOCIOBIOLOGY FALSE?
THE RISE AND FALL OF ISLAM

I come to the final set of criticisms levelled against sociobiology, specifically against human sociobiology. The critics, both those from Boston, and Sahlins, argue that human sociobiology is false. I must confess that, given what has gone before, this in itself strikes me as being a bit of an odd criticism. If sociobiology is unfalsifiable, then I should not have said that it could be shown to be false. Lewontin (1977) has suggested that there is nothing odd here. One can have something ontologically true or false, and yet epistemologically unfalsifiable. However, whilst this may be true of metaphysical or religious statements – it may be false that God is love, and yet unfalsifiable – I am not sure that one can make this case for science. If, as the critics claim, sociobiology is unfalsifiable in the sense that it is immune from attack by empirical counter-evidence, then the critics have no right to turn round and offer empirical evidence as to why sociobiology is false. But no matter: as we have seen, sociobiology is not unfalsifiable in a troublesome sense. Therefore, we can go on to the next question: is it false? Let us consider first the arguments of the Boston critics, and then turn to Sahlins.

Following their accusation of unfalsifiability, the Science for the People critics argue as follows: "There does exist, however, one possibility of tests of such [sociobiological] hypotheses, where they make specific *quantitative* predictions about rates of change of characters in time and about the degree of differentiation between populations of a species." (Allen *et al.*, 1977, p.27, their italics.) In particular, claim the critics, if sociobiology be in any way true, then (restricting our gaze to the human world) we should find major cultural changes are accompanied by (since they are a function of) significant genetic changes. Moreover, we should find significant genetic differences between populations, again reflecting (and causing) major cultural differences.

But, argue the critics, both of these implications prove false. As far as the time dimension is concerned, it is notorious that major cultural changes can and do occur in periods of time far too short for such changes to have been caused by genetic changes, at least by such changes as allowed by orthodox population genetical theory. Thus, for example, the rise and fall of Islam took less than 30 generations, and so that massive upheaval just cannot have been fired directly by the genes. Similarly, speaking for the present but considering matters through space, we find that there is just not the genetic variation between populations required to explain the vast cultural differences. And Lewontin's findings are quoted, namely that 85% of human genetic variation lies within populations, with about 8% between nations and 7% between major races. (Allen *et al.*, 1977, p.28.)

What kinds of replies can sociobiologists make to charges such as these, and how effective are they? There seem to be two basic kinds of reply possible. First, it can be and is indeed admitted that most cultural changes and differences are not essentially functions of the genes. Clearly, the cultural (*i.e.* non-genetic) nature of the Islam case cannot be, and in fact is not, denied. (Wilson, 1976.) But does this not just save sociobiology by gutting it of all content? It certainly has to be admitted that the admission that the rise and fall of something like Islam was not in any way genetic shows that, in certain respects, human sociobiology's claims have severe limits. Nevertheless, the sociobiologists, Wilson in particular, would probably argue that, at least considered across space, the genes might well influence cultural differences, for all that the actual percentages of differences are rather alight. And no doubt, at a point like this, Wilson would invoke his 'multiplier effect', suggesting that small differences at the level of the genotype can explode up into large differences at the level of the phenotype, particularly where behaviour is concerned.

As can be imagined, a counter-move like this does not find much favour with the critics. They believe that Wilson steps out of the frying-pan of falsity straight back into the fire of unfalsifiability. The multiplier effect, and another face-saving effect which the critics find in Wilson's work, the 'threshold effect', according to which organisms have to get to a certain level of complexity in order for the multiplier effect to be able to operate, "are pure inventions of convenience without any evidence to support them. They have been created out of whole cloth to seal off the last loophole through which the theory might have been tested against the real world." (Allen *et al.*, 1977, pp.29–30.)

Are the critics fair in this complaint, or can Wilson properly turn to the multiplier effect to make his case? I must confess that were the whole of (human) sociobiology to rest on this, I would feel a little uncomfortable. Undoubtedly something like the multiplier effect could occur. We are all fully aware that small things can build up into big things. Remember the tale of poor King Richard III.

For want of a nail the shoe was lost,
For want of a shoe the horse was lost,
For want of a horse the rider was lost,
For want of a rider the battle was lost,
For want of a battle the kingdom was lost.
And all for want of a horseshoe nail.

Moreover, it seems plausible to suggest that behaviour stands out above the rest of the phenotype as the element which would most drastically be altered by minimal genetic changes. And also, it does not really seem that the multiplier effect is indeed unfalsifiable. At least, one could surely check it out in other organisms, and if it did not hold at all there one would have a *prima facie* case against its great relevance for humans. Suppose, say, that in mice, small genetic changes had little or no behavioural effects: this would count against the general truth of the multiplier effect.

However, having made these points in favour of the multiplier effect, it must nevertheless be conceded that Wilson's claims for it do strike one as a little on the *ad hoc* side. Wilson hardly provides much by way of strong independent evidence for the importance of the effect: it is introduced to avoid problems. It is true indeed that Wilson suggests that the effect might be important in baboons, where one has two varieties with drastically different behavioural patterns (which are presumably genetically caused rather than just cultural). (Wilson, 1975a, p.11.) Perhaps slight genetic differences here spell major behavioural differences; but as the critics point out, Wilson does not really prove his case, for no real evidence is offered that the baboons are genetically very close. In other words, whilst the critics' charges may not be quite as devastating as they themselves seem to think they are, they are not without some force. Sociobiology has a long way to go on this score.

122   CHAPTER 6

But this now brings me to the second reply that the sociobiologists might make when accused of falsity. They can point out that in important respects the critics' objections miss the sociobiological enterprise entirely. Much of what the sociobiologists want to claim about human behaviour supposedly occurs not only within populations, but also supposedly is repeated from population to population, both in space and in time. In other words, pointing to lack of differences between populations is not very relevant. Pre-literate and advanced industrial societies are both supposed to practice genetically caused reciprocal altruism for the same reasons. Early, middle, and late Islamic populations share the same genes and certain basic behavioural patterns, and nothing that the critics have said proves otherwise.

As far as the Boston critics are concerned, this reply seems both justified and adequate. I would emphasize that at this stage nothing is being said about human sociobiology being true. The point at issue is whether something is being said about human sociobiology being false. And the answer is that appealing to such phenomena as the rise and fall of Islam says nothing about the characteristics common between human beings.

6.7. IS HUMAN SOCIOBIOLOGY FALSE?
THE PROBLEM OF DAUGHTERS

It is at this point that Sahlins steps in, once again. He thinks that basic anthropological data show that all the significant claims that sociobiologists make about genetically caused behavioural constants are quite false. At least, since he restricts his analysis to the purported effects of kin selection, he feels able to support the assertion that "there is not a single system of marriage, post-marital residence, family organization, interpersonal kinship, or common descent in human societies that does not set up a different calculus of relationship and social action than is indicated by the principles of kin selection". (Sahlins, 1976, p.26.) In other words, Sahlins denies that sociobiologists can claim that different societies show similar genetically-caused behavioural patterns.

How does Sahlins try to establish his case? Chiefly through the fate of daughters. Although sons are usually cherished and provided for, in many societies daughters are married off outside the immediate family. This, Sahlins believes, violates kin selection, because an individual ought to be just as concerned with daughters as with sons, and certainly ought not bring in strange women (*i.e.* wives) and care for them rather than one's own blood female kin. He writes:

Take a common rule such as patrilocal residence, with marriage outside the hamlet. By the rule, newly married couples live in the groom's father's household, thus generating an extended family of a man, his wife, his married sons with their spouses and children (family form found among approximately 34 percent of the world's societies, Murdock 1967). By the same rule, the local hamlet — or it could be a territorial hunting band — is comprised of several such families whose heads are usually brothers or sons of brothers. A young man will thus find himself in collaboration with cousins of the first degree ($r = 1/8$) or greater degree ($r = 1/32$, $1/64$, etc.), uncles (FB, $r = 1/4$), quite possibly grand uncles (FFB, $r = 1/8$). If polygyny is practiced there will be even more distant kin within the family (e.g., F1/2BS, $r = 1/16$). Meanwhile, the sister ($r = 1/2$) of this same young man will go off to live with her husband upon marriage, raising her children ($r = 1/4$) in the latter's household; while his mother's sister ($r = 1/4$) has probably always resided elsewhere, as has his paternal aunt ($r = 1/4$) since her marriage. When he grows to maturity, our young man likewise loses his daughter ($r = 1/2$) and her children ($r = 1/4$) as also all other women born to his own extended family group, though he retains his son, his son's son and all males born to the group. Hence insofar as a man favors the 'blood' kinsman of his group, he discriminates against those of equal or closer degree outside of it. (Sahlins, 1976.)

Here, Sahlins believes, we have a flat contradiction with the implications of kin selection. Hence, this example shows that kin selection working on humans cannot be a significant (or insignificant!) factor, at least, not when it comes to the fairly common phenomenon of daughters who leave home after marriage.

Moreover, Sahlins believes that many other anthropological facts show the falsity of sociobiology, inasmuch as it centres on kin selection. For example, in Tahiti it was proper practice to adopt the child of any person whom one had slain in war, despite the fact that owing to the widespread practice of infanticide, one would most likely have destroyed one or more of one's own children. (*Ibid.*, p.49.) And more generally for humans Sahlins believes that the whole kin selection hypothesis collapses because so many people have no ideas of complicated fractions, necessary to work out various kin relations, and because in addition so many have a completely erroneous notion about true biological relationships. "For biologists, the coefficient of relationships among siblings, first cousins and second cousins passes from $1/2$, to $1/8$, to $1/32$, respectively, as compared to the Rangiroan 1, 2, 3. The latter would thus experience some difficulty figuring the egotistic algebra of kin selection posed as a general social logic by the former." (*Ibid.*, p.44.) Indeed, human ignorance about fractions amounts to "a very serious defect in the theory of kin selection". (*Ibid.*, p.45.)

Unfortunately, although Sahlins' objections have an initial plausibility, a little careful thought soon shows that they are nothing like as devastating as he seems to think they are. First, it is regrettable indeed, that Sahlins deliberately restricts himself to kin selection, ignoring reciprocal altruism and parental manipulation. If he had not so circumscribed himself, then he could have seen at once how readily his Tahitian example fits the sociobiological paradigm. On the one hand, reciprocal altruism explains the adoption of enemy's children: if I am ready to do it for you, then you are ready to do it for me. On the other hand, parental manipulation explains the infanticide. By Sahlins' own admission: "In fact, one of the reasons for Tahitian and Hawaiian infanticide, especially among chiefs and other prominent people, appears as an indirect result of the social and reproductive advantages accorded to one child at the expense of his siblings." (*Ibid.*, p.49.) The sociobiologists, themselves, could not have said anything more favourable to their case.

Second, even of the one mechanism which he does discuss, Sahlins shows an almost wanton misunderstanding. As emphasized before, when this kind of objection came up, it is not necessary that fully conscious and accurate assessments of blood-ties be performed for kin selection to operate. Such assessments are not performed in Hymenoptera. They are therefore not needed in *Homo sapiens*. Human kin selection may be false; but pre-literate peoples not knowing fractions does not prove it so. As Alexander writes: ". . . it is not necessary, in inclusive-fitness-maximizing, to *know* who one's kin are, only to *behave* as though one knows." (Alexander 1977, p.12, his italics.)

Third, and perhaps most importantly, Sahlins' whole argument based on patrilocal residence collapses because he forgets sexuality. If human beings just budded off asexually, then one might indeed expect that parents would simply look after their own; but, of course, humans do not bud off in such a way. Humans require mates, and moreover they require mates who are not too closely related: indeed, argue the sociobiologists, so important is it that the mates not be too close genetically, that selection has fashioned our emotions in such a way that we instinctively draw from very close inbreeding. Hence, it is in an individual's reproductive self-interests to find, mate, and live with a non-relative; and it is in an individual's reproductive self-interests to have her or his children find, mate, and live with a non-relative.

So far, so good. But now let us take the argument a little further. Suppose I (a male or a female) mate and bring my spouse home. Suppose my children, for whom I and my spouse have cared, do the same, and so on. I am certainly looking after my genetic self-interests; but I, my spouse, my immediate descendents, and their spouses are doing all the work. Meanwhile, a number of people who also have a direct biological interest in my descendents, namely the parents of my spouse and of my children's spouses, are doing nothing. Kin selection, together perhaps with a bit of reciprocal altruism, suggests that this situation should balance itself out pretty quickly. I will care for half of my children and their spouses, and others will care for the other half of my children and their spouses – they will do so because it is in their direct genetic interests to do so and because if they do not care for others' children, others may not care for their children. (Sahlins himself speaks of the way in which exchanging children helps to form alliances, that is how the care of others' children is a key part of reciprocal altruistic agreements.)

We see, therefore, that the basic structure described by Sahlins is no real problem for sociobiology. Let us now complete the story. Why should it be that sons are kept and daughters are passed on? Well, for a start this is not quite the whole truth. Daughters are cared for and cherished until they are married off, and often they go with a dowry. But there is a possible reason why daughters, not sons, are let go. This derives from the reproductive differences between males and females. Females are much more likely to get pregnant and have offspring than males. (Trivers and Willard, 1974.) Males have to compete for females. This means that one can let females go, because they will almost certainly get impregnated and thus ensure lots of descendents for oneself. Males, however, one has to help to the utmost – for example, by offering family possessions – otherwise they will fail to mate and reproduce. (As mentioned, the dowry perhaps offsets this imbalance.) Consequently, we get males staying home and bringing in spouses. Thus, sociobiological predictions and anthropological findings coincide. Sahlins' attempt to falsify sociobiology fails yet again. As yet, the discipline has not been shown false. (See Hartung, 1976.)

6.8. CONCLUSION

But, is it true? In this, and the previous chapter, I have been concerned to defend sociobiology, including human sociobiology, against the various attacks from without and the various excesses from within. Indeed, some readers must by now be convinced that my commitment to sociobiology is absolute, that I am as enthused about it as any ardent practitioner. This impression, however, is not quite accurate. My feeling is that sociobiology, which I fully admit I find absolutely fascinating, has not been very well served. I think that many of the charges levelled against it have been quite unfair, although it cannot be denied that some of the sociobiologists themselves have said some pretty silly things. They rush in and pronounce confidently on things about which they know very little, and, Mr. Micawber-like, they seem to think that their acknowledgement of problems is enough to make them vanish.

My aim, therefore, has been simply to give sociobiology, including human sociobiology, a chance: not to kill it with criticism or praise before it has really tried to get under way.[3] But already one might think that some sort of progress report is in order. Is there any truth in sociobiology; and, more particularly, is there any truth in human sociobiology? Let us turn in the next chapter to consider these questions.

NOTES TO CHAPTER 6

[1] I do not even deny that modern sociobiologists use some human economic models that Darwin did not: for instance, Oster and Wilson (1978) are quite explicit in their use of such models to explain the ergonomics of insect caste. But first, these models are applied to the social insects, not to humans; second, there are no normative claims; and third, the underlying commitment to traditional Darwinian selection is unshaken, not to say flaunted.

[2] There is considerable debate today as to whether organic characteristics considered at the molecular level generally show adaptive functions. However, fortunately, we can here ignore this controversy, for our concern is with the super-molecular level. (See Lewontin, 1974.)

[3] David Hull (1978) has well compared the present state of human sociobiology, including the attacks, to that of Darwin's theory shortly after the *Origin* was published.

CHAPTER 7

THE POSITIVE EVIDENCE

In order to tackle the problem of the positive evidence for human sociobiology, it might be useful to take a leaf or two out of Charles Darwin's book.
When asked why anyone should accept his theory of evolution through natural selection, he used to reply that there were three reasons. First, there was
what we might call the *direct* evidence from the struggle for existence and the
undoubted variation in the wild. Second, there was the *analogical* evidence,
from artificial selection. And, third, there was the *indirect* evidence, from the
way in which the theory could be applied to so many areas of biological
interest: "In fact the belief in Natural Selection must at present be grounded
entirely on general considerations. (1) On its being a *vera causa*, from the
struggle for existence; and the certain geological fact that species do somehow
change. (2) From the analogy of change under domestication by man's selection. (3) And chiefly from this view connecting under an intelligible point of
view a host of facts." (Darwin, 1887, 3, p.325. See also Ruse 1975.)

Now, without being artificially rigid, let us see where this threefold division carries us in the case of sociobiology. I take it that the problem at issue
is the extent to which human behaviour, specifically social behaviour, is
essentially a function of the genes, where, by this, is meant that the behaviour
will manifest itself under normally occurring environments without the need
for any special learning input. I take it also that the chief rival to a sociobiological explanation of human social behaviour, would be some sort of
'cultural' explanation, where some very definite kind of learning input would
be required. (Because, as we have seen, organic characteristics are all in the
long run the products of both genes and environment, these two explanations
are not necessarily as opposed as just rather implied, and, indeed, we shall be
examining cases where they might run together.)

7.1. THE DIRECT EVIDENCE: PROBLEMS WITH TESTING

By 'direct evidence' of the truth of human sociobiology in this context, I
mean evidence which might be obtained of the genetic control of social behaviour by immediate reference to and experiments upon humans. It is not
difficult to see, unfortunately, that the gathering of such direct evidence –

128   CHAPTER 7

even if it exists – is beset by a number of problems. To bring this point out clearly, consider for a moment the analogous case of gathering direct evidence of genetically controlled social behaviour in a group of lower organisms, for instance the insects.

Obviously, much insect behaviour is tightly under the control of their genes: certainly, if it is not under the control of the genes, it is hard to imagine of what else it could be a function. Paradigmatically, social insects show very complex social behaviour without the need for any learning at all: this is undoubtedly 'genetic' in the sense discussed above of being caused by the genes and developing without any special environmental input. (Wilson, 1971.) However, although I doubt that anyone would seriously question the genetic nature of much insect social behaviour, a little discomfort might be felt about an argument which depends so much on the absence of competing hypotheses. But, assuaging these fears, there is other fairly direct evidence that social behaviour in insects is a function of the genes, namely the fact that behaviour in insects is something which responds to selective pressures (*i.e.* can be intensified or changed through selection). This is something entirely to be expected and understood if behaviour is a function of the genes but quite inexplicable if behaviour be an entirely learned phenomenon. In other words, we have here the opportunity to make direct tests of the genetic bases of insect (social) behaviour.

Perhaps, to underscore this point, it is best to refer to Lewontin's authority here. (I quote Lewontin here because he is an authority; not because it is a matter of controversy.) He writes: "Suppose artificial selection is practiced in a population and succeeds in changing, in a heritable way, the phenotypic distribution in the population. Then it follows that there must have been non-trivial amounts of genetic variation for that character in the population to begin with . . . Even though success in selection does not tell us everything we need to know about genetic variation, it does prove that genetic variation was present to be selected." (Lewontin, 1974, p.87.)

So much for the general point. Now, let us turn to Drosophila. Although these are not social insects, in the sense of having castes and the like, like all animals they exhibit some social behaviour, and we find that this behaviour responds to selection.

Selection for mating preference can be carried out by allowing free mating in a mixture of two mutant stocks and then destroying all hybrid progeny each generation. In this way, for example, Knight, Robertson, and Waddington (1956) changed the mating pattern of *ebony* and *vestigial* mutants from a random one to one in which the ratio of homogametic to heterogametic mating was 1.6 : 1. (Lewontin, 1974, p.90.)

And other aspects of Drosophila close to the sociobiologists' speculations seem subject to selection, for instance whether the Drosophila reproduce sexually or asexually. Thus, in conclusion Lewontin writes: "There appears to be no character — morphogenetic, behavioural, physiological, or cytological — that cannot be selected in Drosophila ... The suggestion is very strong, from the extraordinary variety of possible selection responses, that genetic variation relevant to all aspects of the organism's development and physiology exists in natural populations." (*Ibid.*, p.92.) In short, selection experiments suggest that, like other characteristics, insect social behaviour is genetically controlled.

But, what bearing does this all have on human social behaviour? Obviously, in dealing with our own species, quite apart from the fact that the hypothesis that human behaviour is essentially cultural or learned cannot be dismissed or ignored as readily as in the insect case, we have at least two special problems: First, one has the simple problem that humans are relatively slow-breeding. This means that (even if one wanted to) one cannot easily carry out such selection experiments as Lewontin and others have carried out on Drosophila. Even if one were prepared to select for something like (to take a fairly uncontroversial matter) sporting ability, it would take several hundred years before one could hope to have any significant results. Second, there are the moral problems. The idea of deliberately forcing breeding programmes upon people strike most of us as pretty horrendous. It would be bad enough selecting for sporting abilities; selecting for something like criminal tendencies would be truly dreadful (although for devotees of horror movies it does rather fire the imagination: "Jack the Ripper, meet Madaleine Smith.").

How then can we try to get around both the practical and the moral problems and tackle the difficulty posed by the fact that much human social behaviour can *prima facie* as readily be explained through culture as through the genes? We need some way of separating out the environment from the heredity, and to do this, fairly obviously what we need to do is to look for natural experiments: that is to say we need to look for things which have in fact occurred, which simulate experiments which one might like to have performed. In this way, both the temporal and the moral difficulties get reduced. Ideally, I suppose, what one would like would be records of deliberate human breeding programmes (so one could slough off responsibility on others!) However, even if they exist, these are few and far between. But there are many other pertinent examples of natural experiments. For instance, to take a non-behavioural example, we are all fairly certain that skin-colour is essentially a genetic rather than a cultural phenomenon, because we know that no matter what climate people are brought up in, within certain limits they have the skin colour of those surrounding them. Similarly, to take a behavioural example, we are all fairly certain that speaking English rather than French is essentially non-genetic, because the language one speaks most naturally seems virtually entirely a function of one's developmental environment, specifically whether one is surrounded by English speakers or by French speakers.

But with the appropriate kind of approach necessary, broadly delimited, matters start to get a little more tricky, particularly when it comes to subtleties of behaviour. Is a child a good sports player because the parents are good sports players and they all share the same causative genes? Or is a child a good sports player because the parents are good sports players and they all share the same causative environment? In some few cases, we can say fairly definitely that behaviour is genetically caused because the behaviour is associated with chromosome irregularities and often also with typical phenotypic traits (which latter could not be a function of culture). Best known of these cases is Down's syndrome, caused by an extra chromosome, and leading to mental retardation as well as various physical signs. (Bodmer and Cavelli-Sforza, 1976.) But most cases of behaviour are not this easily traced to their origins. I very much doubt that the causes of Pele's abilities show up under a microscope. Here, as in most other cases of behaviour, the dilemma of distinguishing environment from genes remains; and the only way out of it seems to be through the study of natural experiments involving relatives, where, for various reasons, genetic and environmental, the influence of the genes and the influence of learning can be partially or wholly torn apart.

Probably the most famous of such natural experiments are studied in so-called 'twin tests'. (Shields, 1962; Bodmer and Cavalli-Sforza, 1976; Vandenberg, 1976.) There are two kinds of twins: monozygotic twins who share exactly the same genotype, and dizygotic twins who do not share the same genotype, and who therefore have exactly the same relationship as any pair of siblings (i.e. 50%). Normally speaking, both sets of twins will have been brought up together as children, that is they will have shared the same environment. Hence, should monozygotic twins be behaviourally closer then dizygotic twins, one can put down this closeness to the genes, rather than the environment. In other words, here indeed one does have an opportunity of separating out the genes from the environment.[1] Furthermore, one can obviously (at least in theory) ring some changes on this basic pattern. For instance, if one can find some examples of twins brought up apart, that is, not sharing the same environment, then this gives one additional ways of separating the effects of the environment and the genes.

It should be added that, although twin tests are the best-known way of separating out human environmental and genetic causes, in fact one is not restricted exclusively to studies of twins, or even of siblings, because parents and children are related to each other as well! One can, for example, hope to separate out the effects of the genes and the environment if one has detailed information on children who have been adopted. Do such children seem more like their natural parents (*i.e.* are the genes in control), or do they seem more like their adoptive parents (*i.e.* is the environment in control)?

We see therefore that studies of natural human experiments give us hope of ferreting out possible genetic causes of human social behaviour. We must turn next to the results.

7.2. SUCCESSES AND RESERVATIONS

Dating back from the work of Francis Galton (1869), Charles Darwin's cousin, who asked whether famous men tend to be related to other famous men, many studies of the kinds just illustrated have been made, and all sorts of aspects of human behaviour have been examined. For instance, several studies have been made on schizophrenia, the most common human mental ailment. Moreover, these studies show that schizophrenia most certainly has a significant genetic factor. "Familial correlations are relatively high, with the incidence in sibs of the affected being of the order of 12 percent, that in dizygous cotwins of the same magnitude, while monozygous cotwins of affected have an incidence of at least 40 to 50 percent." (Bodmer and Cavelli-Sforza, 1976, p.516.) The only reasonable explanation of these findings is that the far higher correlation for monozygotic twins is caused by their genetic identity (although since the correlation is not perfect, clearly schizophrenia has some environmental causal components too.)

Similarly to schizophrenia, the second most common human mental ailment, manic depression, probably also has a genetic component. Although there is still speculation about the causes, the inheritance pattern suggests that it may be a sex-linked dominant, and indeed manic depression seems linked with other sex-linked genes, such as those for colour blindness. (*Ibid.*) And as for these mental ailments, many other suggestions have been made about possible genetic bases of human behaviour, which suggestions have then borne some fruit. For example, there is some evidence that alcoholism may be partly a function of the genes. (McClearn and Defries, 1973.)

From studies of human natural experiments, what I am including under the heading of the 'direct evidence' for human sociobiology, we can therefore definitely assert that there is ground for belief that some human behaviour or forms of such behaviour is controlled by the genes. I should add that some of the evidence is sufficiently unambiguous and strong that even the critics of sociobiology are prepared to allow its relevance and force. The Boston critics, for instance, agree that schizophrenia has a genetic component. Furthermore, it is hard to imagine that none of this genetically caused behaviour would have any fairly significant social implications. Not all schizophrenics have had quite the effects that Joan of Arc had (assuming that she was one); but when one person in a hundred is affected this way it is bound to have some ramifications. Of course, this is not to say that the sociobiologists and their critics would agree on the exact nature and magnitude of these ramifications.

Unfortunately, however, we cannot just leave matters at this point. At least two serious reservations must be made to studies of natural human experiments, inasmuch as they pertain to the truth of sociobiology. First, there is the obvious point that for all that these studies may prove that human social behaviour can be affected significantly by genetic causes, this is far from saying that human social behaviour will be as much under the control of the genes as some of the sociobiologists rather imply. Also, it is not to say that all of the things that sociobiologists specifically single out are under the control of the genes.

For instance, following up on this last comment, I am not aware that any studies have been made to show that human altruism is under the control of the genes. (Trivers, 1971, in fact, is quite open about the lack of direct evidence.) Nor incidentally is it that easy to see how such studies might be made, for apart from the difficulties of defining precisely what one might mean by 'altruism' in a particular case, it is crucial to the success of such studies that we be dealing with behaviour that some people possess and other people do not possess. Otherwise, it is impossible to sort out the effects of the genes from the effects of the environment. (Consider: if everyone exhibits behaviour $\emptyset$, then monozygotic twins show it, dizygotic twins show it, natural parents show it, adoptive parents show it, and so forth.) But one of the most distinctive features of the sociobiological thesis about human altruism is how widespread it is. Remember how Trivers saw reciprocal altruism in all kinds of society.

Hence, on the one hand, as a matter of fact, many of the features of human social behaviour that sociobiologists think might be under the control of the genes have not as yet been studied directly, and one suspects that in some cases it will be some time, if ever, before they are studied. I suppose it is logically possible that one might do a study on incestuous desires, but the practical difficulties seem enormous: were I sleeping with my sister (which I am not!), the last thing I should do would be to admit it on some prying geneticist's questionnaire.[2] And, on the other hand, as a matter of logic, many of the features of human social behaviour thought under the control of the genes, specifically those thought possessed universally, are precluded from direct studies of the type being discussed. (Note: I am not saying that this all makes it unreasonable to believe the sociobiologists. What I am saying is that the direct evidence is limited.)

The second point of reservation that ought to be made here concerns well-publicised doubts about the authenticity and reliability of so many of the studies which have been done on, what I am calling, human natural experiments. I believe that, despite all the criticisms, the studies have shown that some human behaviour is genetically controlled. But it would be dishonest not to admit that some grandiose claims have been made for the power of such studies and that many such claims have withered, or at least wilted, under the force of well-justified criticism.

Consider, for example, one of the studies on which Wilson builds his case, namely Kallman's (1952) study purportedly showing that there is a genetic causal factor influencing homosexual behaviour. *Prima facie* this study strongly underscores the genetic component in homosexuality, for of 85 sets of twins studied by Kallman (where at least one twin was homosexual), in all 40 monozygotic cases both twins were homosexuals and showed virtually similar intensities of such behaviour, whereas in the dizygotic cases most of the cotwins of distinct homosexuals showed no homosexual behaviour at all. However, against this study and its conclusion we have the fact that other studies have found significant correlations between male homosexuality and age of mother at birth (older mothers have more homosexual sons), and between male homosexuality and birth order (younger sons have more tendency to be homosexual). One could perhaps explain these findings as being functions of mutations: just as in the case of Down's syndrome, older mothers are more likely to carry an ovum mutated to cause homosexual behaviour. But there is no independent evidence backing this supposition, and obviously there are rival environmental explanations to account for the findings. (Pare, 1965; Marmor, 1965; Rainer, 1976.)

Moreover, specifically relating to Kallman's work, yet other studies have found monozygotic twins discordant for homosexuality. Also, Kallman was undoubtedly biased towards finding genetic factors influencing behaviour. A study by him purportedly showing a genetic basis for schizophrenia was clearly influenced by his expectations. Hence, the same might be true of his homosexuality study. And furthermore, non-genetic explanations for Kallman's findings are possible. For instance, C. W. Wahl has suggested that because of the close identification of identical twins this might break down the incest taboo between them thus leading to freer sexual activity between them (members of the same sex!), which could then get transformed into full homosexual behaviour. (Pare, 1965.) Overall therefore, whilst these points certainly do not rule out Kallman's conclusions — Wahl's suggestion particularly strikes me as being *ad hoc* — they make one recognize that it would be unwise to assume without doubt or qualification that homosexuality has a significant genetic basis.

More generally speaking, I am certainly not claiming that all studies on human behaviour suffer from problems like these just listed for Kallman's study; but, without doubt, there are many which do. Hence, for reasons like these, as well as for the reservations made earlier, we must recognize the somewhat limited support that human natural experiment studies give to human sociobiology.

7.3. THE QUESTION OF INTELLIGENCE

Although I have admitted to reservations about human behavioural genetical studies, at this point the reader may complain that, perhaps out of cowardice, I am skirting some of the most important issues pertaining to studies of the bases of human behaviour. Moreover, under a facade of reasonableness, I am failing to mention some of the most serious reservations of all that must be drawn about such studies: reservations which have a very significant impact on the truth-value of the human sociobiological programme. In the eyes of such a reader, whilst it is no doubt interesting to find that schizophrenia is a function of the genes, and, given what the sociobiologists have had to say about homosexuality, certainly pertinent to raise the question of its genetic basis, in an important sense these facts and questions are all a bit peripheral. If what the sociobiologists claim is true, such a reader will argue, then what we ought to find is that some of the really important behavioural characteristics or capacities are controlled by the genes. Otherwise the human sociobiological thesis is rather trivial.

Now, putting the matter this way, the reader might continue, one behavioural ability or capacity stands out above all others in importance: intelligence. Hence, if human sociobiology is saying anything interesting, it ought to be including a claim that intelligence is in some way a function of the genes. But now triumphantly concludes this reader, if she or he suddenly is revealed as a not-very-friendly reader, we see the threadbare nature of the direct evidence of sociobiology, for studies purportedly proving that intelligence is a function of the genes are notoriously unreliable — fraudulent even. Hence, inasmuch as human sociobiology depends on direct evidence it is a house built on sand.

I must confess that I personally would be inclined to deny that if something affects up to 10% of the population, the kinds of figures cited for homosexual behaviour, it is all that peripheral. But let us leave this point and turn to the question of intelligence. The objection just given obviously depends on one being able to draw a tight link between human sociobiology and claims that human intelligence is strongly causally influenced by the genes: a link the rightness of which might well be questionned. Nevertheless, it is undeniable that human sociobiology, as presently constituted, and claims about genetic bases of intelligence (or something related) are connected. Wilson, for example, openly states that ". . . the influence of genetic factors toward the assumption of certain *broad* roles cannot be discounted". (Wilson, 1975a, p.555.) It will be appropriate therefore to conclude our look at the direct evidence for human sociobiology by considering very briefly the troubled question of intelligence and its causes. I shall begin by turning to recent work on the matter and then see how it relates to sociobiology.

7.4. THE CAUSES BEHIND INTELLIGENCE

Although the debate about the causes of intelligence goes back at least to Plato, who assumed in his *Republic* that intelligence has heritable causes and that therefore one ought set up breeding programmes to produce Philosopher-kings (although all the emphasis on education in the *Republic* shows that Plato thought there were other causes too), as might be expected it was not until this century and the development of psychology as an independent science that intelligence causal studies with any real appearance of scientific solidity started to appear. Possibly the most impressive of all these studies were those of the noted British psychologist, Sir Cyril Burt, who supposedly traced a large number of twins separated at birth, and, who, on the basis of these studies, concluded that the genes play a very significant role in determining intelligence. (Monozygotic twins had much closer IQ scores than did dizygotic twins). (See, for example, Burt, 1966.)

Unfortunately, however, since Burt's death (in 1971) grave doubt has been cast on the authenticity of his work. (Wade, 1976b.) And indeed, although there is still debate about whether Burt deliberately concocted or falsified data, it is generally agreed that his studies are essentially worthless. It goes almost without saying therefore, that this makes far more hypothetical the many claims about human intelligence which have been based on or draw upon his studies. One thinks here particularly of two highly controversial recent theses. First, that of Arthur Jensen, who pointed to the fact that peoples of different races in the U.S. score differently on IQ tests (specially, whites score higher than blacks), and who suggested that this difference is best explicable in terms of intelligence being controlled by the genes and of different races having different genes. (Jensen, 1969, 1972.) Second, that of Richard Herrnstein, who argued that socioeconomic standing is a function of the genes, and that, consequently, as we move towards a meritocracy we move towards genetically distinct layers or castes within societies. (Herrnstein, 1971.)

Of course, even if Burt was a fraud this does not as such make false the claims of people like Jensen and Herrnstein, nor obviously does it make generally invalid studies of human natural experiments, even studies designed to find the causes of intelligence. But, keeping the discussion at a general level and avoiding the fascinating side-track of Burt's personal ethic of research, it does not take much imagination to see that there are grave problems ahead for those who would study causes of intelligence, including the most open and honest of investigators. For a start, there is the problem of precisely what one means by 'intelligence' and how one measures it. With some reason, there have been a plethora of critics who have objected to normal methods of measurement, namely IQ tests. They object that, at best, such tests gauge ability to do well on IQ tests! (Block and Dworkin, 1974; Kamin, 1974.) And even if we lay aside objections like these, formidable methodological and logical barriers still stand in the way of would-be investigators. Hopes of success and the kinds of problems can perhaps best be illustrated by reference to a recent study done by the psychologist, Harry Munsinger (1975a): a study which by some has been hailed as a worthy replacement for Burt's work. (Herrnstein, 1975.)

In order to test the extent to which IQ could be inherited, Munsinger studied 41 children (20 Mexican-American and 21 Anglo-American) who were adopted very shortly after birth. From information contained in adoption records, Munsinger found that the correlation between education-social status of adopting parents and children's IQ (Lorge-Thorndike scores) was very low (−0.140), whereas correlation between socio-economic status of biological parents and children's IQ was very high (+0.700). Naturally, therefore, Munsinger felt that he had helped support the case for genetic control of intelligence. ". . . the most reasonable conclusion from all these data seems to be that biological parents exert a significant effect on both the average level and the rank order of their children's intelligence, even when separated at birth, and that, by contrast, adopting parents have little influence on either the average level or the relative ranking of their adopted children's intelligence scores." (Munsinger, 1975a, p.254. See also Munsinger, 1975b.)

However, this confident conclusion has been attacked by one of the most thorough critics of claims that intelligence is genetically controlled, the Princeton psychologist Leon Kamin. (1977a, 1977b. See also 1974.) On the one hand, Kamin objects to the fact that Munsinger had to work indirectly through other people. Because the state of California (where Munsinger did his study) will not let outsiders study adoption records, Munsinger had to get state officials to gather his data, which gathering involved such things as making estimates of parents' socio-economic status. On the other hand, Kamin objects to some of the values that Munsinger had to read into his data to obtain his conclusions. For instance, Munsinger had to set up a scale for what would count as socio-economic *cum* intellectual rank. (*E.g.* college graduate score 1, grade school drop-out rank 6.) Objects Kamin:

[Munsinger's procedure] contains an obvious artifact. The determination of whether the child's intellectual level is 'more similar' to that of biological or adoptive midparent depends on the arbitrary numerical values assigned to levels of education, in relation to the equally arbitrary scale value assigned to an IQ of 100. For example, consider a child with an IQ of 111 whose biological parents each completed only 3 years of school but whose adoptive parents were a college graduate and a college dropout. With Munsinger's scales, the child's IQ rank is 4, the biological midparent intellectual level is 6, and the adoptive midparent intellectual level is 1.5. The child, according to Munsinger, more closely 'resembles' the biological parents in "intellectual level." (Kamin, 1977b, p.412.)

Hence, concludes Kamin, with the kind of bitterness which the reader must be concluding is the norm in science:

I can only reiterate that Munsinger's study is rife with error, and is invalid. When a scientist indicates that he 'cannot report precisely' how his data were obtained, he has nothing to report – and less than nothing if those data fall into 'unusual distributions'. This, I continue to think, should have been obvious to the editors of a responsible scientific journal. (*Ibid.*)

I give this example of Munsinger's study and Kamin's attacks both to show the present-day status of studies on the genetics of IQ and to show the grave problems which still surround the whole topic. (Munsinger, 1975b, contains a good review of the literature, and discussion of problems.) In fact, for myself,

I am inclined to think that Kamin's objections are not quite as devastating as he himself thinks. It seems plausible to suggest that even an official of the California State Adoption Agency could assess people's socio-economic status, given clear guide-lines. And I suspect that the extreme cases as Kamin hypothesises even out in the whole study.

I am uncomfortably aware that one swallow does not a summer make, and that one study, even if defensible against criticism, does not prove unequivocally that the genes play a significant causal role in human intellectual abilities. I am, however, writing a book on the sociobiology controversy, not the IQ controversy, despite the fact that the two overlap. Having, therefore, given the reader a taste of the positive evidence which has been produced in favour of the genetic thesis, I shall now have to take a leap beyond what I am able to offer here, referring those of my readers who still have open minds on the subject to the pertinent literature, and draw what seems to me to be the most reasonable conclusion on the matter. This, somewhat hesitantly is that, if one considers all the pertinent studies which have been made, some case can be made for the claim that at least some elements of intelligence are under control of the genes. I realise that the justifiable attacks on Burt's integrity have left IQ studies under somewhat of a cloud; but then after all we do not reject Mendelian genetics, even though the results Mendel cited to support his case are far too detailed to be true. (Fisher, 1936; Wright, 1966.)

Analogously, I believe there are enough sound, untainted results to support at least a limited claim about the importance of the genes in intelligence, or, more precisely, that different people have different intelligences in part because of different genes. I find it hard to imagine that there is not some pertinent genetic difference between the members of the Sociobiology Study Group of Science for the People and (say) a random group of farm labourers. An obvious and necessary qualification to a claim like this about the heritability of intelligence is that this is all more a question of capacities to respond to certain learning situations, rather than of abilities that will develop willy-nilly. In other words, the environment is a crucial factor in intelligence development, whatever the genetic factors may be. (See Jencks et al., 1972.)

In order to keep the discussion moving, let us now grant, even if only hypothetically, that intelligence, roughly understood as the sort of thing which shows up on IQ tests, is to some degree under the control of the genes. What implications does this have for human sociobiology? Clearly, we are still a long way from establishing an extreme sociobiological case: a case more extreme I would add than today's sociobiologists seem to want to establish. Consider a claim like Jensen's, namely that the genes make peoples of some races on average brighter than peoples of other races (specifically, whites over blacks), and a claim which I have argued strenuously I do not find in the work of Wilson and his fellows. It simply is not the case that because intelligence may in some way be a function of the genes, blacks are therefore intellectually inferior to whites, even though blacks may score lower on IQ tests than whites.

Allowing that IQ tests do have something to do with intelligence, this conclusion still does not follow for at least two reasons. First, because IQ tests are notoriously biased in favour of whites: in an extreme case, children were shown pictures of beautiful white women and ugly black women and asked to choose the prettiest! Admittedly, the extreme cases may no longer exist, but still it is hard to deny that more subtle forms of cultural discrimination may yet exist. (See Block and Dworkin, 1974.) Second, because no one is denying the influence of the environment on intelligence, or denying that if people are deliberately forced into a certain situation or held in that situation, this can affect their intelligence. That blacks may score lower on IQ tests is readily explicable by the fact that many grow up in culturally deprived circumstances and, moreover, have no way in which they are able to break out.

Furthermore, in addition to these two reasons showing that one cannot exclude the environment as the main causal factor yielding IQ differences between groups, it might be mentioned that there are empirical studies supporting the suspicion that the environment plays a crucial causal role behind the IQ differences between groups. One famous U.S. study showed that black conscripts scored lower as a whole than white conscripts as a whole, but that Northern blacks scored higher than Southern whites: a phenomenon most readily explicable by the better living standards in the North. And another well-known study on Scottish schoolchildren confirms this conclusion. Over a number of years successive classes of children scored better on IQ tests. This could not have been a function of change of genes, but most certainly could have been a function of improved education. Perhaps a good parallel in this context is with height, something known definitely to have both genetic and environmental factors, and which like IQ increased in Scottish schoolchildren over the years. We know that height increased in this case because of environmental changes, namely improved nutrition; analogously, it is hard to imagine that environmental changes were not also responsible for the jump in IQ. (Bodmer and Cavelli-Sforza, 1976; Dobzhansky, 1962.)

Turning now from race to roles, a connexion between intelligence and the genes probably has more signification for differences in role performance than it has for racial differences in performance. And indeed, as we have seen, the sociobiologists would want to make some claims here: "... the influence of genetic factors toward the assumption of certain *broad* roles cannot be discounted." (Wilson, 1975a, p.555.) Nevertheless, note what is not being claimed and what in fact cannot be claimed from a connexion between intelligence and the genes, namely that the roles we play in society are strictly determined by our genes. It is true that success on IQ tests and professional success are correlated: people with IQ's of 70 tend not to be doctors or even philosophy professors! Hence, in societies where people really do have freedom of choice, one might certainly expect the genes to have some influence on one's role, or, more precisely, that different genes would influence people towards different roles. But the environment obviously also has an effect on the roles one plays, and in addition there are clearly many factors other than intelligence which influence one's falling into roles — perseverence, ambition, stamina, and so forth. Hence, there is no exact isomorphism between genes causing intelligence and social roles. Interestingly, confirming this point but at the same time perhaps confirming the importance of genes in our social roles, there is an inverse relationship between schizophrenia and social status. Since schizophrenia often does have a genetic causal component, this points towards the influence of the genes on roles. (Cancro, 1976.)

Furthermore, to admit some connexion between a heritable intelligence and roles is not to concede an extreme position like that of Herrnstein. There is sufficient social (and sexual!) intercourse between members of different groups in human societies to ensure a healthy flow of genes in all directions. Perhaps the most we can say is the modest conclusion of Wilson:

> The heredity factors of human success are strongly polygenic and form a long list, only a few of which have been measured. IQ constitutes only one subset of the components of intelligence. Less tangible but equally important qualities are creativity, entrepreneurship, drive, and mental stamina. Let us assume that the genes contributing to these qualities are scattered over many chromosomes. Assume further that some of the traits are uncorrelated or even negatively correlated. Under these circumstances only the most intense forms of disruptive selection could result in the formation of stable ensembles of genes. A much more likely circumstance is the one that apparently prevails: the maintenance of a large amount of genetic diversity within societies and the loose correlation of some of the genetically determined traits with success. (Wilson, 1975a, p.555.)

I have said enough on the subject of the genes and intelligence: my critics will no doubt say that I have said more than enough. But as I bring this stage of the discussion to an end, it is perhaps not inappropriate to remind the reader that because something is the case, it does not imply that it ought to be so. Even if what Wilson writes just above is roughly true, it does not imply that we could never find ways of manipulating the environment to increase abilities, or that we ought not practice this. As the parent of a child with a reading disability, I am all in favour of such environmental manipulations. Perhaps some day we shall indeed find ways to turn children with IQ's of 70 into doctors and philosophy professors; nothing which has been said rules out this possibility, although the morality of so doing is another matter!

7.5. THE WEIGHT OF THE DIRECT EVIDENCE FOR HUMAN SOCIOBIOLOGY

We have not covered all of the areas in which studies have been or are being made of possible genetic influences on human behaviour: deliberately I have decided that the problem of intelligence raises quite enough controversy and hence have stayed away from the equally heated question of whether genes can cause criminality. (McClearn and DeFries, 1973.) Nevertheless, we have now a good idea of the strengths and weaknesses of tests for the direct evidence of human sociobiology. And we have also an idea of the results obtained. It is undoubtedly the case that some aspects of human behaviour are under control of the genes and that these could well have significant social effects. Nevertheless, it is far from obvious just how much human behaviour is controlled by the genes or precisely what the social effects are of that behaviour which is so controlled.

As far as some of the claims that sociobiologists make about human behaviour, probably their confidence outstrips their evidence: perhaps some of their claims about homosexuality fall into this category. Conversely, as far as some of the more modest sociobiological claims are concerned, it may well be that there is truth in what the sociobiologists say: perhaps the claim that genes partially influence our roles is a good example here. But, as pointed out, the direct evidence is scanty or altogether missing for some of the more interesting human sociobiological claims, for instance about incest or altruism. Of course, the direct evidence is not the only potential evidence for human sociobiology, and so following Darwin's order, let us turn next to the question of the analogical evidence that might be adduced in favour of human sociobiology.

7.6. THE ARGUMENT FROM ANALOGY

The problem to be discussed here is, with respect to the genetic bases of behaviour, to what extent one can argue legitimately from the animal world

to the human world? We enter here a matter of some controversy and difference between sociobiologists and their critics, so first a couple of general
comments about analogy might not be amiss. First, perhaps obvious but
nevertheless worth saying since the critics of human sociobiology seem frequently to forget it: analogy *per se* is not a bad argument. Indeed, analogy
can be a very good argument and it is certainly indispensible. I need to buy a
pair of shoes, and I therefore decide to buy a kind of a make which I bought
before because they wore so well. This is an analogical argument, and it is just
good common sense. The second point, again obvious but worth saying since
the sociobiologists seem frequently to forget it: analogy must be used with
care or it can go wrong. If I decide to buy a pair of shoes because they are
the same colour as a previous pair which wore well, then I am using an analogy but not a very good one.

What are the criteria of 'goodness' for an analogy? This is not an easy
question to answer, but the *relevance* of the properties being invoked seems
crucial. (Salmon, 1973, pp.97–100.) Consider: In an analogy one goes from
one thing to another, arguing that because the two things share a number
of properties (say *a, b, c*) and because the first thing has a certain other
property (say *d*), it is reasonable to suppose that the second thing has that
property also. But two objects always have some properties in common,
and indeed they always have some properties not in common. What is
crucial is how relevant these various properties are to the inference in question. Inasmuch as one has relevant similar properties one has a stronger
analogy and inasmuch as one has relevant dissimilar properties one has a
weaker analogy. Thus, in the case of shoes, the make that they are seems
to be a relevant property with respect to quality whereas colour does not.
Of course, with respect to other ends, say coordination with one's clothes,
the reverse might be the case, and fairly obviously much disagreement about
analogical arguments revolves around whether certain properties in certain
situations are relevant or not. It should be pointed out that making analogical
arguments can be well worthwhile, even when for various reasons one is
not altogether convinced of the truth of the conclusion. Such arguments
can be fruitful sources of new hypotheses and insights, which hypotheses
and insights one might then try to confirm independently. (Ruse, 1973b,
1973c.)

Now, turning to sociobiology, the general scheme of the analogical argument seems as follows: animals and humans share many biological attributes,
particularly in the way that genes cause morphological characteristics. Moreover, these attributes are relevant to the causes of behaviour. Animals have

genetically caused behaviour. These are no relevant dissimilarities great enough to rule out an inference to genetic bases of human behaviour. Therefore, it is reasonable to conclude by analogy that there are genetic bases for at least some human behaviour. Moreover, when animal behaviour and human behaviour are very similar in other respects, and when there is good reason to suppose the animal behaviour controlled by the genes, there is reason to conclude that the human behaviour is also controlled by the genes.

Now, taking matters at a very general level, I suspect that few would want to deny this argument completely. Human beings feel sexual attraction, they mate, and they look after their children. Given our animality and its crucial role in our continued existence, there is little doubt that our genes play a significant role in the arousals that the members of one sex feel because of the bodies of the other sex, or in the care parents show their children. (As will be made very clear later in the chapter, I am *not* here arguing that everything in sex and parenthood is a function of the genes.)

The major question at issue is how much further one can go beyond very vague generalities like these. The sociobiologists think that one can go quite a bit further, and it is worth quoting Wilson on this:

Characters that shift from species to species or genus to genus are the most labile. We cannot safely extrapolate them from the cercopithecoid monkeys and apes to man. In the primates, these labile qualities include group size, group cohesiveness, openness of the group to others, involvement of the male in parental care, attention structure, and the intensity and form of territorial defense. Characters are considered conservative if they remain constant at the level of the taxanomic family or throughout the order Primates, and they are the ones most likely to have persisted in relatively unaltered form into the evolution of *Homo*. These conservative traits include aggressive dominance systems, with males generally dominant over females, scaling in the intensity of responses, especially during aggressive interactions, intensive and prolonged maternal care, with a pronounced degree of socialization in the young; and matrilineal social organization. This classification of behavioural traits offers an appropriate basis for hypothesis formation. It allows a qualitative assessment of the probabilities that various behavioral traits have persisted into modern *Homo sapiens*. (Wilson, 1975a, p.551.)

I suspect that this in itself would have been fairly controversial, but then Wilson rather muddies things by adding the following:

The possibility of course remains that some labile traits are homologous between man and, say, the chimpanzee. And, conversely, some traits conservative throughout the rest of the primates might nevertheless have changed during the origin of man. Furthermore, the assessment is not meant to imply that conservative traits are more genetic − that is, have higher heritability − than labile ones. Lability can be based wholly on genetic differences between species or populations within species. (*Ibid*.)

As might be expected, the critics have castigated Wilson at this point, accusing him of inconsistency and of trying to have matters every way at once. If traits are conservative then we have evidence of genetically caused traits in humans; if traits are labile then we have evidence of genetically caused traits in humans. *Ergo*, conservative or labile, we have evidence of genetically caused traits in humans. (Allen *et al.*, 1977.) However, whilst in the above-quoted passages Wilson is certainly not as clear as one might wish, I think one can save him from the charge of inconsistency and make some sense of what he is saying. What he does not seem to be claiming (at least, what he ought not to be claiming) is that by analogy from animals one can argue that all human behaviour is genetic. Wilson is accepting that the cultural dimension for humans is a relevant dissimilarity; although, obviously, he would argue also that the dissimilarity is not so great as to rule out any genetic influence on human behaviour.

What Wilson does, in fact, seem to be claiming are three things. First, when animals (specifically primates) show very similar behaviour it is reasonable to suppose that it might have a significant genetic component. The reason for this is because culture leads to all kinds of variation. This is, as Wilson says, not to assert that conservative traits are necessarily more genetic than any labile traits. It is however to assert that in the case of conservative traits we can be more certain of genetic causes than for labile traits, because for labile traits culture is more likely to have been a major factor. Second, Wilson is also claiming that given conservative animal traits and their probability of being genetic, and given humans' having the same traits, we can argue analogically that the human traits are genetic. Third, Wilson is pointing out that there are relevant dissimilarities lurking, and so we should not think we have proved more than we have!

No one, I take it, is going to deny the third point, and I would suggest that it is not unreasonable to accept the first. It is, of course, true that as we come up through the higher animals cultural factors become more significant. Wilson himself details the case of Imo, the genius Japanese macaque, who discovered two radically new feeding methods, which discoveries were then transmitted by learning through her group. (Wilson, 1975a, p.170.) But this is in no way to deny the fact that animal behaviour is influenced by the genes, and that a telling sign that cultural factors are at work is that we have rapid change and variability: a major reason why we do not think the English language genetic is because the French do not speak it and because the Anglo-Saxons did not either. In other words, if we have no change across primate species, because cultural causes would predispose us to believe that there would be change, it seems reasonable to think that there may be significant genetic factors.

This now brings us to the second point, the key point. Can we argue from the primates to humans? The relevant dissimilarity is the far greater human cultural realm. The purported crucial relevant similarities are, apart from morphological similarities, primate behaviour patterns (now presumed) genetic and human behaviour patterns. The point about culture is indisputable. The important question is whether the positive properties can be sufficiently relevant and sufficiently similar to overcome the negative aspect. My own feeling is that, in principle, they can be; although, as Wilson admits, one needs to accept one's conclusions with some caution. But relying on the conclusion drawn above, namely that if culture is a major moving factor we might expect to see at least some differences, if we find fairly complex behaviour shared generally by humans and by the primates, then I believe a case has been made for regarding the human behaviour as having a significant genetic component. The dissimilarity between animal and human behaviour is breeched at this point.

I must nevertheless hasten to add that I am making a theoretical case. I am not at once endorsing all of Wilson's claims about particular traits. Take, for example, his claim about males being dominant over females. In fact, I am far from convinced in this case that Wilson has offered enough evidence of the similarity between humans and other primates: indeed, he himself admits that male dominance does not always hold in the primates, and as I have pointed out before, I am not sure that it always absolutely holds in humans either. (Remember the lament of the wretched Mr. Bumble upon being told that in law he was a power over his wife and thus responsible for her actions. "If the law supposes that . . . the law is a ass — a idiot. If that's the eye of the law, the law is a bachelor.") Later in this chapter, I shall return to the question of male—female differences.

I argue, therefore, that in theory one can argue analogically from the primates to humans; but in practice I am far from convinced that the sociobiologists have yet provided enough evidence to do so, at least in any definitive way. And the same conclusions hold even more strongly as we move towards animals farther from *Homo sapiens*. The greater the gap between the brutes and humans, the less reliable arguments by analogy become. I certainly would not want to rule them out *a priori*, but, except in the very broadest outlines, such inferences strike me as being at this stage in the development of human sociobiology, more of the nature of fruitful heuristic suggestions than providers of well-established conclusions.

7.7. HUMAN AGGRESSION

The reader might feel that this is a rather weak conclusion that I am drawing about analogies from the animal to the human sphere. I would not want to pretend that it is that strong; but understood for what it is worth, it can, I believe, be a useful tool. To illustrate this point, let us refer back for a moment to Maynard Smith's (1972) discussion of aggression. Unlike the American sociobiologists, he denies explicitly that he has said anything applicable to humans. Even though at one point explicitly he applies his ideas to primates and even though at another point he makes reference to neurotic humans, Maynard Smith wants to argue that the only analogy between humans in a game-theoretic situation and aggressive animals is a formal one: one of logical structure. And the implication of his argument seems to be that similarity of logical structure is not a sufficiently strong similarity, or perhaps not even a relevant similarity, to support any inference about the possibility of human and animal aggressions having roughly similar biological foundations. Nevertheless, I suspect that a proper understanding of the nature and power of analogy shows that one can draw a less restricted conclusion than this.

For a start, as a number of philosophers have pointed out, many supposed formal analogies in science turn out to depend crucially on material analogy too, that is on analogy between the terms of concepts of the two situations. (For example, Achinstein, 1968.) Thus, for example, although water flowing through a tube and electricity flowing through a wire share formal structure, the material analogy of something going through a long object is crucial. Without it we would not find the analogy at all inspiring, as we do not find inspiring the analogy between flowing water and the many other phenomena which share the same formal equations (*e.g.* change of electrostatic charge on an oil drop).

Whether material analogy is necessary in all significant scientific analogies is a matter we can ignore here. (See Hempel, 1965.) Of importance to us is, first, the fact that in many of the most fruitful scientific analogies amongst the most relevant similarities have been the material analogies (indeed, because we have material analogy we are tempted to look for or posit formal similarity), and, second, there is the obvious fact that, whatever Maynard Smith may say, there is indisputable material analogy between two people engaged in a ritualized conflict (say, boxing according to the Queensbury rules) and two animals engaged in restrained aggression. Apart from anything else, one has two sets of organisms contesting. Therefore, I am far from convinced that, inasmuch as they prove successful in the animal world, it is improper to take Maynard Smith's theories and apply them back to the human world.

The logic of Maynard Smith's argument seems to be that there are sufficient pertinent or relevant similarities between humans involved in certain sorts of conflicts and animals involved in certain sorts of conflicts, that we can plausibly take formal analyses worked out in the former cases and apply them to the latter cases. And whatever Maynard Smith may say he is doing, without doubt some (if not all) of the similarities which he is taking as relevant here are material factors, such as that in both cases we have organisms pitted against each other. By his own admission, he first got involved with game theory because he knew that it was to do with 'conflicts', and this was his area of concern in the biological world. (Maynard Smith, 1972, p.13.) But analogies work two ways. If A is like B, then B is like A. Hence, inasmuch as human aggressive behaviour simulates animal aggressive behaviour, and inasmuch as Maynard Smith's ideas work for animals, one can try to apply them back to humans. Or, putting the matter another way, if human conflicts contain sufficient relevant similarities to animal conflicts to try transferring formal analyses from one area to another, then they contain sufficient relevant similarities to try transferring the analyses back again.

Of course, I am not pretending that animals consciously think out strategies as humans often do, and that thus there are total similarities, formal and material, between the human and animal worlds. But, there are more ways than one of skinning a cat, or less metaphorically, evolution often finds different causal routes to achieve the same ends. Thus, even though we may recognize significant differences between animal and human aggression, this is not to deny that the same broad analysis may be applicable to both. And, in any case, one suspects that much human aggression is not always that rationally thought out anyway, so the analogy between animals and humans need not flounder just because of consciousness. In short, I see no reason to exclude aggression from the scope of human sociobiology, or to deny the relevance to humans of work on the topic done so far exclusively with reference to animals.

Do not misunderstand me. I am certainly not suggesting that at this point one can simply conclude that vital aspects of human aggression are adequately explicable by means of models such as Maynard Smith's. Indeed, I do not as yet think that this is true of animal aggression. But what I do think is that analogy makes it worthwhile to see where the application of Maynard Smith's models to the human world leads us, and that consequently inasmuch as someone like Alexander (1971) suggests an analysis of human aggression which bears marked similarities to the approach of Maynard Smith in the animal world, he can rightly feel that Maynard Smith's work offers analogical support for the kind of approach that he takes.

7.8. THE INDIRECT EVIDENCE FOR ANIMAL SOCIOBIOLOGY

We come now to the third, and final, possible source of support for human sociobiology: the kind of evidence that one has for a theory when it leads to true predictions or implications. A good example of this kind of evidence, what I am calling the 'indirect evidence', would be the pertinence of the fossil record for Charles Darwin's theory of evolution through natural selection: the record, a kind of branching progression from general to specific, from embryonic to adult forms, was just the kind of record one would expect given Darwin's theory.[2] I take it that a highly relevant consideration here is the extent to which a theory explains or predicts better than its competitors: in Darwin's case, the fossil record gave severe difficulties to the various transcendentalist and Christian progressive revelatory rival hypotheses. (Bowler, 1976b.) As pointed out in a previous chapter, evidence of this kind may never make a theory absolutely true, but it is both crucial and sometimes very persuasive indeed: particularly if a theory implies surprising or hitherto-thought-false true phenomena.

As in the case of the direct evidence, it is useful to start with the animal world and then, by contrast, turn to the human situation. With respect to the animals, summing up much of earlier discussion, the following facts seem both true and relevant. First, mainly due to intensive studies in recent years, we have gained and are still gaining a great deal of knowledge about animal behaviour, and we are realizing that much of this behaviour is in some sense of the term 'social'. Thus, in the case of aggression, we have various kinds of hostile but restrained encounters between conspecifics. In the case of sexuality, often we have elaborate interactions between males and females, for instance protracted courtship displays and struggles between males for possession of harems of females. In the case of generation differences, there is often lengthy and exhausting parental care of offspring by members of one or both sexes. And then on top of all of this we have various kinds of 'altruism', between relatives, as in the case of Hymenoptera, and between strangers, as in the case of cleaner and cleaned fish. Nevertheless, we are learning also that animal interactions are not all sweetness and light. Aggression sometimes escalates to all-out fatal conflict. Sexual struggles are also sometimes carried to the death of one contestant, as well as spelling death for any infant who unfortunately happens to get in the way. There is conflict at weaning time between parents and offspring. And finally, altruism sometimes breaks down or gets subverted.

The second point is that there are a number of possible alternative explanations of this social behaviour. One could, I suppose, suggest some sort of cultural explanation, hypothesizing that much animal social behaviour is learned or similarly transmitted. One could certainly invoke a group selection hypothesis, arguing that the behaviour must be understood in terms of genes which are of adaptive advantage to the group, most probably the species. Possibly one might explain much of the behaviour in terms of random factors, arguing that generally there is no systematic reason for animal social behaviour. And finally, one can obviously explain the behaviour in the ways that the sociobiologists do, namely in terms of various kinds of individual selection mechanisms.

Third, it is fairly clear that almost all of these explanations will not do. Undoubtedly, some animal behaviour, particularly in the higher vertebrates, is learned or cultural. In this chapter, I have already mentioned the case of Imo, who first discovered how to wash potatoes and then how to separate wheat and sand by throwing both on the sea (the wheat floats): this behaviour then spreading by learning throughout the group. (Wilson, 1975a, p.170.) Nevertheless, it is obvious that most animal social behaviour is not learned, that is, is not cultural. There is, for instance, no way in which the Hymenoptera could learn all of the intricate social acts that they perform. Similarly, except perhaps in very rare cases, a group selection hypothesis fails. We have seen earlier some of the internal difficulties with group selection. In addition, it is refuted by the breakdowns in intra-specific social behaviour. *Contra* to what people like Lorenz claim, animals do sometimes kill their fellow species' members, and so forth. Finally, although possibly some animal social behaviour is random, in the sense of not being related to any systematic cause, it is hard to see that essentially all is. This goes against our whole understanding of evolution in a general sense, and almost by definition leaves inexplicable some phenomena which seem to cry out for explanation: why, for instance, should castes have evolved independently so often in the Hymenoptera, but only once elsewhere?

The fourth, and concluding, point is that animal sociobiology can explain many of the phenomena of animal social behaviour, including here many of the awkward pieces of data. Trivers' (1972) analysis of sexuality, for example, gives one understanding of various forms of male—female interaction, and the same is true of his analysis (1974) of parent—offspring relations: conflict between parent and child at time of weaning is seen, not as some random, unfortunate side-effect of evolution, but as the result of definite selective pressures ultimately causing different, conflicting behaviours. And similarly, many other aspects of animal social behaviour are explicable by sociobiology, in ways discussed earlier.

We see, therefore, the various points which pertain to animal sociobiology; and the logic of the overall argument is obvious. Sociobiology and only socio-biology makes reasonable claim to explaining the facts of animal social behav-iour, and it does this in a fairly unifying sort of way. Therefore, it, and it alone, can lay reasonable claim to our support. Although, as we know full well, at the present this support must be hedged with qualifications and reser-vations. In the case of something like aggression, what we have are plausible possibility models, not hard quantitative predictions. And even where it looked at one time as though we were getting strong predictions, predictions of the best kind because they were certainly not of very obvious or expected phenomena, serious doubts have now been raised. I refer, of course, to Trivers and Hare's (1976) derivations of Hymenoptera sex ratios, and to Alexander's objections to their work. (Alexander and Sherman, 1977.)

Rome was not built in a day and animal sociobiology cannot be proven overnight. With respect to indirect evidence, a good start has been made. Let us leave it at that and turn now to human sociobiology.

7.9. THE INDIRECT EVIDENCE FOR HUMAN SOCIOBIOLOGY

I have no reason to think that the logic of testing theories about humans differs from the logic of testing theories about animals. As for animals, there-fore, let us run through the pertinent facts and see where this gets us.

First, there is the question of the phenomena. Almost all human behaviour is social, in the sense that we are using this term. Perhaps masturbation is an example of non-social behaviour, although even this has a role in promoting certain kinds of bonds. Moreover, in many respects, human social behaviour is phenomenologically analogous to social behaviour in the animal world. Thus, for example, we get aggression amongst humans, particularly when resources like food and space get very limited. Also, undoubtedly, we get restrained aggression amongst humans, involving bluff and threat, but some-times escalating into all-out violence. As we all know, perhaps only too well, we have marked sexual dimorphisms, with males larger and stronger than females (although usually less long lived) and often (although by no means always) in socially dominant roles. And we certainly get a great deal of ritual in intra-sexual relationships, with males traditionally taking more aggressive leads and so forth.

With respect to parenthood, both sexes traditionally invest a great deal of time and effort, and there are frequently times of tension as children reach the final stages of growing up. Finally, for all that they are often very selfish, humans do show wide forms of altruistic behaviour, both to relatives and non-relatives. Moreover, it does seem roughly true to say that the closer someone is related to us, the more we let them presume on us without hope of return. Some people are indeed saints; but most of us when dealing with strangers are prepared to do significant things for them, only because we expect returns of one sort or another. (This latter is not necessarily a crude demand of a tip for favours rendered, but could, say, be a salary for teaching others' children.)

The second point to be made is again, for the human case, not very different from the animal case. As for animals, we have a number of possible explanations of human social behaviour. Such behaviour could be cultural, in the sense that we are understanding it, namely learned or similarly transmitted: something which goes from phenotype to phenotype without having to be encoded in some way in the genes. It could be a result of a biological group selective force, in which case such behaviour would be encoded in the genes and of biological advantage to the group rather than to the individual. The behaviour might be a function of random factors, perhaps explicable in each particular case but certainly with no systematic reasons. And then of course, human sociobiology, concentrating as it does on individual selective forces, might be a totally adequate explanation of human social behaviour, both as it has developed and as it is.

We come now to the third and fourth points, and it is here obviously that we start to veer from the animal situation. I take it that we can dismiss, without too much question, the group selection hypothesis and the random factors explanation. I am not saying that in the human case there would be absolutely no truth in either, but as attempts to explain anything like the whole gamut of human social experience they seem inadequate. There is too much intra-human strife for group selection, and if one rules out learning *and* the genes, it just does not seem plausible that there would be so many regularities between humans (say) with respect to long-term mating and child rearing. One is denying that humans are animals and that they are humans! This then leaves us with the cultural explanation and the sociobiological explanation.

Reversing the order somewhat, and mentioning what corresponds to the fourth point of the previous section, I am going to assume here that a case has been made for saying that, with some reservations, human sociobiology can account for human social behaviour. In an earlier chapter, I tried to show this from a positive viewpoint, and then in the last chapter particularly I have tried to show this from a negative viewpoint by defusing supposed factual counter-examples, particularly those anthropological ones put forward by Sahlins. Sociobiology may not be true; but I do not think that the critics have shown that it is not.

But the trouble is, of course, coming back to the third point, namely that concerning the truth or falsity of the rivals of sociobiology, so much that the sociobiologists want to explain in terms of individual selection working on the genes, can also be explained in terms of a fairly obvious cultural model, involving discovery, learning, and so forth. Moreover, unlike so much of animal social behaviour, one cannot just dismiss the cultural explanation, because it is so clearly true that culture could have brought about at least a large part of human social behaviour. Thus, let us just run through the dimensions of human social behaviour which we have been discussing, trying to put the case for cultural causes as strongly as possible: in fact, stating matters in a way that we might expect a critic of sociobiology to state them.

7.10. THE PLAUSIBILITY OF CULTURAL CAUSES OVER BIOLOGICAL CAUSES

In almost all societies, from the most primitive pre-literate to our own (that is not necessarily an order of merit!), aggression, fighting, war, and so forth, are a significant part of culture, broadly defined. Moreover, there are lots of ways in which this aspect of culture could be self-perpetuating. For example, we learn about fighting from books, films, and television: John Wayne, a man who makes his living pretending to kill his fellows, is one of the folk heroes of our time. And it is perhaps not without significance that when Wilson wanted examples of humans turning aggressive under strain, he turned to fiction. (Wilson, 1975a, p.255.) In other words, there is no difficulty in finding cultural causes for aggression.

Turning next to sexuality, our critic will say, again we find that *prima facie* culture is a significant influence. Even at the most basic level, this seems true. Consider, for example, the Western male's obsession with female breasts, yearning for the sight of a nipple, things in open display in so many non-Western societies. It is hard to imagine that what separates us from the Africans on this is anything but a function of learning, customs, and so on. In a like manner, one doubts that what separates us from the Victorians in not finding the sight of an ankle all *that* erotic is anything other than culture. Similarly, at a more general level, it is hard to deny that learning could be responsible for males thinking they ought to be dominant and so many females thinking that they ought to be dominated. And the same holds for something like incest taboos. The elders of the tribe, having seen that close inbreeding has horrendous effects, thereupon ban it for the good of everyone involved. (See Harris, 1971.)

Undoubtedly, our critic will feel that just the same points hold for parenthood and for altruism. Much of the ability (or lack of ability) we show in raising children is learned, whether it be from our own parents or from Dr. Spock. And, certainly, there is no shortage of literature telling us about how to deal with parent—children conflicts. Nor is there a dearth of good advice when it comes to altruism, not to mention all sorts of other cultural influences, explicit and implicit, on our behaviour at this point. We are taught all manner of things about how we ought to behave towards others, through the church, through education, and through informal pressures from family and friends. Even when it comes to something like the mother's brother phenomenon, of which Alexander makes so much, cultural explanations spring readily to mind. All one needs is for some people to realise that their wives' children are liable not to be their own, and presumably this is not that implausible in societies which are postulated to have systematic and widespread extra-marital sexual intercourse, and the phenomenon could appear: then being passed on by custom and learning.

Furthermore, it is not that difficult to think up cultural hypotheses to explain behaviour, perhaps today not so obviously altruistic, for which the sociobiologists are prepared to invoke their various kinds of altruism-producing mechanisms. Consider, for example, homosexual behaviour, to account for which the sociobiologists suggest that kin selection may have been important (homosexuals, being freed from raising children of their own, therefore turn to raising the children of close relatives). One can just as readily suggest that homosexuality is a function of various kinds of learning experiences. In Ancient Greece, for instance, homosexual love was consciously held up as an ideal; and perhaps even today homosexuals have the sexual orientation that they have because of various subtle and not-so-subtle pressures on them as they grow up. English public schools have a notorious reputation for producing homosexuals, and this is readily explicable in terms of the fact that during a key period of sexual development boys are kept locked away from female influence. For what it is worth, A. S. Neill, founder of the world-famous free school, Summerhill, where boys and girls mixed freely, boasted that he had never produced an active homosexual. (Neill, 1960.)

In short, the critic will argue, considering matters with respect to indirect evidence, in the human case as opposed to the animal case, the cultural explanation of social behaviour cannot be readily dismissed. Rivalling the sociobiological explanations of human social behaviour, we have very plausible cultural explanations. Furthermore, it might be felt also by such a critic that, if only on a principle of simplicity, in a great many cases the cultural explanations ought to get the nod over the sociobiological explanations. In almost all spheres of human social behaviour, we know that cultural influences do exist. Why then argue that essentially these have no causal influence, that they are, as it were, unimportant scum on human existence, and that really human behaviour is controlled by unknown and unseen genes? We know full well that people, particularly young people, are subjected to a barrage of propaganda about such things as aggression and sexuality, and that, by and large, they react as though they are influenced by it. Why, therefore, deny a real causal influence here and instead argue that it is hypothetical genes which are the real motive forces?

Moreover, it might be suggested also that as one descends to some of the details of human social behaviour, although sociobiology can indeed provide explanations, they start to look increasingly *ad hoc*. Consider, just for an example, a change which is taking place in our society. In recent years, certainly in the last fifty, we have seen a dramatic increase in divorce. This has meant, amongst other things, that a great many children are no longer living with their biological parents (particularly fathers), but with step-parents. Consequently, even though the biological parents may not lose contact with their children, and even though the step-parents may often not show quite the full concern for their step-children that they would for their own children, many people today are in fact showing much parental concern for children that they know are not their biological offspring. Now, as a sociobiologist one might argue, since obviously this change in patterns of child-care cannot be a function of change of genes, that what we have now are genes for reciprocal altruism coming into play: I look after someone else's children, knowing that someone will look after my children, and so forth. But (objects our critic), why bother to get into all of this when one can invoke such obvious cultural influences, for instance the decline of religion (brought on by the rise of science), thus leading to a devaluation of the sanctity of marriage, and so forth? Why reject culture in favour of hypothesized biology?

Or take again homosexuality. Why raise the whole subject of kin selection when there are possible environmental causes readily at hand? Particularly since, if one does rely on a kin selection hypothesis, then necessarily one is committed to the view that homosexuals reproduce less than heterosexuals. Now this may indeed be true of males, although, as our critic will probably point out, the sociobiologists seem not to have any hard figures on this matter, and the Kinsey report certainly suggests that many homosexuals have heterosexual experiences. But, as our critic will remind us, there is also the problem of female homosexuals. Are we to assume that the same genes cause female homosexuality as those which cause male homosexuality? If not, then we must start hypothesizing yet more genes. But if the same genes supposedly causing male homosexuality also cause lesbianism, then we must square this fact with the other sociobiological speculation that, by and large, females do not have much choice about reproduction. (Trivers and Willard, 1973.) That is that, even if a woman has lesbian yearnings, she is as liable to reproduce as anyone. In other words, whatever a woman's sexual inclinations, her biological fitness is not affected. No doubt, our critic will conclude somewhat sneeringly, this fact too can be squared with kin selection, but by this stage one might start to wonder if it is all worth the effort.

At this point, probably, the real critics of sociobiology will be rubbing their hands with glee. Having defended human sociobiology against so many criticisms, through the mouthpiece of my imaginary critic, I seem now just to be conceding everything to the other side. We have cultural explanations of human social behaviour, and we have sociobiological explanations of such behaviour. And what I seem to be suggesting is that, if only by an application of Occam's Razor, it makes most sense normally to side with the cultural explanations. Even if human sociobiology be not yet proven wrong, it is not yet proven true; nor at present is there good reason to adopt it as the most plausible hypothesis. However, for a number of reasons I think this is too extreme a conclusion to draw. I do not believe that the time has come to write *finis* to human sociobiology.

7.11.   DOES CULTURE LEAVE A PLACE FOR HUMAN SOCIOBIOLOGY?

First, it must be pointed out that at this stage of my argument I am restricting myself to human sociobiology as it relates to its indirect evidence. Even if one believes that human social behaviour in the broad sense can be explained by cultural mechanisms at least as well as by sociobiological mechanisms, one might still not want to reject sociobiology at once because of other reasons, for instance the direct evidence introduced earlier in this chapter. If, for instance, one does take seriously the notion that intelligence has a strong genetic component, then one is, at a minimum, setting limitations on the effect that culture can have on the roles we all play in society. (That is, the effect that culture would normally have. Consciously aware of our biology in this matter, we might try to get around it.) Or suppose one accepts Kallmann's twin-data on homosexuality. Then, although it may seem all rather forced and *ad hoc* to get into involved explanations about kin selection, this does not necessarily mean that it is wrong to do so. As pointed out earlier, a cultural explanation of monozygotic-twin homosexuality in terms of the relaxing of incest taboos, seems about as *ad hoc* as it is possible to be. In other words, the direct evidence, not to mention the analogical evidence, might incline one to persevere with human sociobiological explanations.

The second point, one made already in our discussion of analogy, is that however pervasive cultural influences may be, it is hard to deny (at the very least) that there is a basic biological structure on which it is founded. As also noted in the discussion of analogy, perhaps the thing which distinguishes culture most from biology is that the former can and does change much more rapidly than the latter can and does. Cultures can change almost overnight, certainly in less than a century, whereas biology is much more slow-moving, requiring thousands of years. Now if culture does have all this flexibility then one might expect that if everything is culture, some societies would have broken drastically away from others. But, although some peoples may be fairly peaceable, it is questionable about how many groups show absolutely no aggression when faced with severe limitations on resources. Also, with respect to sex and parenthood, there are certainly variations, but apart from isolated temporary societies, like that of the Shakers, humans generally do mate and get involved in long-term child care. Frankly, if having children is all a matter of culture (or, at least, if keeping them is), then given the effort that raising them entails, I am surprised that the childless family was not invented long before the wheel. And, even in the case of altruism, there seem to be some constants. Consequently, even if one wants to give culture as wide a scope of action as possible, whether humans are entirely blank sheets may be questioned. Could one bring humans up to have absolutely no aggressive tendencies, no interest in sex, no feeling for children, and no willingness at all to relate altruistically towards others? This is what an extreme culturalist position would seem to imply.

Third, it might perhaps be objected that stating the case for culture against biology almost entirely on the basis of the West is a little unfair. No sociobiologist is going to deny that culture can take us away from our biology, or that in the past few thousand years, but a drop in the evolutionary bucket, humans have started to explode upwards in a cultural fashion. What is true and interesting in sociobiology, it might thus be argued, is that in our very recent past we humans were subject to biological forces, that, consequently, we have genes fashioned by these forces which, although now essentially masked, might break out in some ways, and that less industrialized peoples still let their biology show through in significant ways.

I must confess that I am myself a little dubious about this defence of human sociobiology, although no doubt it has some truth in it. Certainly sociobiologists do believe that culture can mask many aspects of our biology, and probably in the future it will be possible to mask even more. Nevertheless, sociobiologists certainly give the impression that as things stand at the moment, many aspects of Western social behaviour relate directly to human biology. Trivers, for instance, freely uses Western examples in his discussions of sex, parenthood, and altruism. Consequently, whilst I suspect that in many respects the sociobiologists think Western society atypical, it is not considered *that* atypical. Hence, using examples based on the West is not really that unfair, even though it may be thought that it is in less developed societies that sociobiology has its most direct relevance.

Fourth, it could certainly be pointed out that at one end of the scale human behaviour starts to slide into morphology and physiology, or at least to get very intimately connected with these. Thus, for instance, even if one does not much like the sociobiological explanation of why women stop menstruating in their forties, whereas men remain fertile much longer (Alexander, 1974), it is hard to see how a cultural explanation could have any relevance at all. And the same holds true of the fact that unfit women tend to have a far higher ratio of female children than is normal. (Trivers and Willard, 1973.) But if culture fails us at this point whereas sociobiology does provide plausible hypotheses, then it must also be remembered that hypotheses like these are not just isolated suggestions but fairly direct consequences of fundamental sociobiological principles — in these cases just mentioned about child care and sexual differences. Hence, perhaps in instances like these we have another reason for taking sociobiology seriously.

In this context, of course, it is hard not to wonder about the whole sociobiological approach to male—female differences. Physically female and male humans are very different, and, moreover, to an evolutionist it is difficult to imagine that some of these differences are not a function of selection. Women's broad hips, for example, seem clearly to be an adaptation for childbearing: more particularly, for the bearing of children with the very large heads that human infants have. One might even go so far as to say that given the physical differences between male and female humans, it is difficult to imagine that some of these differences are not a function of the kinds of selective forces hypothesized by the sociobiologists. The larger physical size of males over females, for instance, is clearly not a function of culture, and since it is a phenomenon fairly common in the animal world too, and since the sociobiologists seem to be on the right, or at least a hopeful, track towards explaining it in the animal world (and explaining the exceptions as in fish), there seems a natural presumption that the sociobiological explanations might to this extent be applicable in the human world also.

For this reason, to pick up a point left dangling from earlier chapters, I would suggest that there is sufficient *prima facie* physical evidence and success in other fields to justify a sociobiological approach to female—male human differences, without thereby incurring a label of 'sexist'. Something like Freud's analysis of female psycho-sexual development does strike me as being sexist, at least as judged from today's viewpoint, because there appears to be little or no evidence that the empirical facts so crucial to this analysis have any real existence. It seems to me highly improbable that little girls do in fact desperately covet their brothers' penises. (Freud, 1905.) On the other hand, there are physical differences between males and females, and drawing attention to them and trying to explain them as do the sociobiologists is not necessarily to read in the anti-female values which underlie sexism.

Matters cannot quite be left at this point. What I have just been saying applies directly to sociobiological explanations of human sexual physiological differences. When it comes to actual behaviour, as I have already made clear, I think the sociobiologists are on much shakier ground. With respect to human male dominance, I would be inclined to accept that it is biological in the secondary sense that it is women who have babies, get more encumbered by them than males and so forth, and thus tend to fall into the less active roles, from an outward perspective. If one likes, one can say that they have the biological capacity to accept a male-dominated culture. But I am not sure that facts or theory imply anything stronger: for instance that, because of biology, women necessarily crave domination by men, or that, if freed from childbearing as happens increasingly today, women could not take on hitherto-exclusively male roles. To argue this seems to me to go beyond the evidence. I do not think such implications are necessary for human sociobiology, and when individual sociobiologists get close to making them as Wilson does once or twice, I catch a whiff of sexism.

Continuing now with our list of points suggesting that, as yet, it is premature on the basis of the indirect evidence to discard sociobiological causes for cultural causes, we have the fifth point that, even if one is unenthused about human sociobiology, it is surely still a little premature to drop it entirely before more effort has been made to test it. Take, for example, the mother's brother phenomenon, upon which Alexander stakes so much of his case. Is it indeed true that it is virtually all and only those societies where paternity is regularly in doubt that exhibit the phenomenon? Also, are there some of these societies where a cultural explanation really just does not seem plausible? For instance, in some of these societies are people so unaware of the mechanics of the reproductive process that males might genuinely be unaware that they are not the biological fathers of their social children; or are people sometimes quite incapable of consciously working out true blood relationships, yet instinctively able to sense what it is in their reproductive interests to do? Conversely, do we ever find societies where the mother's brother phenomenon obtains, and yet social fathers are usually always biological fathers too? Much more careful study seems to be in order here.

In this context, however, it should be added that already Alexander thinks that there is strong evidence for his cousin hypothesis (that parallel- and cross-cousins are distinguished when parallel-cousins are liable to be half-siblings). He writes:

Considering the extremes, then asymmetry in cousin treatment should be concentrated in societies favoring or specifying sororal polygyny, and asymmetry should be concentrated in societies practicing monogamy. Almost half (211) of the 423 societies in Murdock's (1967) ethnographic sample (of 565), usable for this purpose because the relevant data are there, treat parallel- and cross-cousins symmetrically or do not distinguish them, and half (212) treat them asymmetrically or distinguish them. But 75 of 79 societies (95%) favouring or prescribing sororal polygyny treat parallel- and cross-cousins asymmetrically, while only 35 of 101 monogamous societies (35%) do so ($p < 0.0001$). Alternatively, using Murdock's (1967) standard sample of 186 societies, substituting where data are not available for a few societies indicated in the sample, one finds that only 5 of 15 monogamous societies (33%) treat cousins asymmetrically, while 7 of 8 sororally polygynous societies (87.5%) treat them asymmetrically ($p = 0.0177$: Fisher's exact probability test; Siegel, 1965). (Alexander, 1977b.)

Even if one continued to demand further evidence or independent test, it is clear that these are early days at present on the basis of the indirect evidence to conclude that sociobiology is not worth taking seriously as a plausible explanation of human social behaviour: before there has even been extensive testing of its success as compared to cultural explanations of such behaviour.

The sixth and final point is perhaps the most important, not the least because it may show us a way out of the whole sociobiological controversy. It therefore merits careful introduction.

7.12. A BIOLOGICAL-CULTURAL COMPROMISE

The discussion up to now has surely been based on a false, at least artificially rigid, dichotomy. I have presented the choice of explanations of human social behaviour as being one of either culture or biology (specifically the genes): if behaviour is to be explained culturally then it cannot be explained biologically, and if it is to be explained biologically then it cannot be explained culturally. But somehow this all seems rather implausible. On the one hand, as pointed out, it seems incredible to suggest that human biology has no causal effect at all on human social behaviour. Just think what our behaviour would be like if women, like many mammals, came into heat. On the other hand, we must surely agree with the hypothetical critic that it seems equally incredible to suggest that human culture has no causal effect at all on human social behaviour: that culture is mere epiphenomenal froth on the top of the biology. Could it not, therefore, be reasonably suggested that the true causes of human social behaviour lie not in the genes alone nor in invention and learning alone, but in some amalgam of the two?

The natural question which now arises is precisely how biology and culture might mix causally? One possibility might be that biology is responsible for some behaviour and culture for other behaviour, but essentially both work separately. However, whilst there may be some truth in this, one doubts that it is all or even most of the truth. Consider an already-introduced example, the erotic feelings that someone like myself, a Western male, experiences at the sight of a naked female body. Surely my biology has some part in these feelings: I do not get anything like the same sensations at the sight of a naked tree. On the other hand, my culture has some part in these feelings: breasts excite me in a way that they do not excite people of other cultures, and ankles do not excite me in a way that they have excited peoples of other times. In other words, biology and culture seem to come together causally in some sexual social behaviour: and just as for sex, one suspects that a similar case could be made for aggression, parenthood, and altruism.

But there is still the problem of finding exactly how biology and culture come together to produce a particular item of social behaviour. An obvious suggestion is that the genes set the limits to a wide range of possible behaviours, the genes, as it were, give us certain capacities, and then learning and so forth instantiates the variables and determines the particular behaviours. Human behaviour can thus be seen as biologically adaptive, which is what the sociobiologists want, but crucially causally influenced by learning, which is what the culturalists want. The whole learning process thus is seen as introducing a fantastically powerful mechanism for causing adaptive changes in the phenotype at a rate far in advance of anything possible if one relies solely on pertinent changes in the genes.

To me, as to a number of other thinkers, a compromise like this has many attractions. (Dobzhansky, 1962; Ruse, 1974; Campbell, 1975; Durham, 1978.) It seems natural to accept that, overall, my response to females is adaptive, even though the details of my response are not explicitly coded in my genes. One does not, of course, have to deny that some behaviours will be more strictly determined by the genes than others, or conversely that humans have evolved to the point where deliberately or subconsciously they can now break from much of their biology, in the sense that through culture they might adopt many behavioural practices which are not, in fact, that adaptive. However, embracing a position like this undoubtedly one will, in fact, see biological constraints on much human behaviour, particularly when one turns to pre-literate societies, and similarly one will be looking generally for some biologically adaptive significances to human social practices, or at least for special reasons as to why they might not be there. (Let me emphasize once again, however, that this is not to say that one cannot decide deliberately to break from the biologically adaptive.)

Clearly, the kind of compromise position just sketched must be argued for, not just postulated. *Prima facie*, one doubts that it would be that acceptable to someone like Sahlins, who, as we know, has been at pains to deny that human mating practices do have biologically adaptive values. In the next chapter, therefore, I shall be introducing some work which suggests at least that it is a proposal worth pursuing. But it can be noted here that probably some of the sociobiologists themselves would not react that unfavourably to the proposal. Thus, for example, Alexander writes as follows about the possible way in which the mother's brother phenomenon might operate:

In regard to mother's brother, there is little doubt that fathers in the relevant societies have fewer and less satisfactory social interactions with both their wives and their wives' offspring than in other societies where mother's brother is not prominent. Men are thus in a position to *learn* not only to behave toward spouse's offspring as if they are unlikely to be their own true offspring, but also to behave as though realizing, even if not actually realizing, that sister's offspring not only represent a reasonable alternative investment, but also are in need because of their mother's spouse's failure to care for them ... In other words, the cultural change represented by mother's brother becoming prominent in some societies could have resulted from the changes in learning situations provided in the two societies, leading to different circumstances that in each case result in individual-fitness-maximizing by the parties involved in the changes. (Alexander, 1977a, p.19, his italics.)

What we seem to have here is something very much in tune with what has been talked about in this section.

Enough now has been said by way of preliminary about how biological and cultural factors might come together. Let us leave these matters, at least until the next chapter. And indeed, with this introduction of a possible resolution to some of the harsher aspects of the sociobiological controversy, let us now bring to an end the discussion of the indirect evidence for human sociobiology.

7.13. CONCLUSION

In three different ways, we have looked for evidence supporting human sociobiology: directly, analogically, and indirectly. We have cast our nets out and they have certainly not come up empty. If the arguments and data of this chapter are at all sound, then in all three of our directions we find at least some reason for taking human sociobiology seriously. From direct experiments, including here natural experiments, there is evidence that some important human social attributes may well have significant genetic backing. From analogy, at the least we get valuable heuristic guides and suggestions. And as we have just seen, the indirect evidence also gives us reason to think that there may be a future for human sociobiology; particularly if serious effort is made to blend together the best parts of both the biological and the cultural explanations of human social behaviour.

Thus, for an example which nicely ties everything together, consider for a moment the whole question of incest taboos, something which has always fascinated anthropologists and upon which the sociobiologists feel that they can throw explanatory light, namely explaining them in terms of biological urges under the control of the genes. (Alexander, 1977b; Barash, 1977; Parker, 1976; Van de Berghe and Barash, 1977; Wilson, 1977b. But see also Harris, 1971.) Although, as I have pointed out, there is no really practicable question of supporting the taboos' possible biological basis through direct evidence of the kind which involves studies of the actual inclinations of descendants of incestuous unions, with respect to all three areas of possible confirmation there is some evidence positively inclined towards the sociobiologists' hypothesis. First, directly, there is the evidence that children of the kibbutz rarely, if ever, want to marry (or have intercourse with) those with whom they were raised. They feel no genuine sexual desires towards those who were socially their siblings, even though they know full well that there are no biological links between them, nor are there religious, legal, or social barriers to their unions. In other words, the direct evidence suggests that human biology (as opposed to culture exclusively) makes one unable psychologically to relate sexually to those with whom one is raised. Clearly, this has implications for incest taboos, for normally, social siblings are biological siblings. Second, there is analogical evidence of biological bases for incest taboos: simply, some of the higher primates avoid very close inbreeding, even when the opportunity presents itself. Third, most obviously there is the indirect evidence that except in the very rarest of cases, human incest taboos holds rigidly and universally, and that biologically speaking such taboos are highly adaptive. The effects of close inbreeding are dreadful. (Adams and Neel, 1967.) In short, there seems all round evidence for the biological basis of such taboos.

Nevertheless, even if one grants some strength to a case like this, speaking generally it would obviously be foolish and misleading to pretend that at present human sociobiology can lay claim to being anything like a well-established scientific theory. Even in the animal world sociobiology has far to go, and the path to be trodden is that much farther in the human world. We have a lot of speculations: fascinating speculations undoubtedly, but speculations nevertheless. (Ruse, 1977c.) It would therefore be less than candid were one not to concede that there are occasions when the sociobiologists let their enthusiasms outrun their evidence. Indeed, there are times when the evidence is so thin that the reader must wonder why I have been at such pains in this book to defend human sociobiology. This, however, is to misunderstand my intention. I have not set out to write an apology for human sociobiology, arguing that the reader ought to accept it in its entirety. Rather, my driving force has come from the fact that I believe that there is no more important study than that of humankind, that I believe also that humans are animals, and that, consequently, I have concluded that it would be downright silly not to explore to the fullest the possible implications of human animality on our social behaviour: particularly given that this is a time when, at long last, biologists are really coming to grips, phenomenologically and theoretically, with animal social behaviour.

My reading of the history of science is that when important new scientific theories are introduced, their audacity frequently far outstrips their hard evidence. If they are really worthwhile then they will bear fruit: they will lead to new discoveries, methods of testing, unifications, and so forth. Otherwise, for all of the initial enthusiasm, like second-rate best sellers they will fade away, soon to be forgotten.[4] My own feeling is that human sociobiology, especially if it can combine with cultural forces, has, in fact, got an important future. Apart from anything else, I do not see it as an entirely brand new theory, but as an exciting extension of a well-established theory, neo-Darwinian evolutionary theory, into the realm of human social behaviour. Hence, I do not think it stands entirely alone (as perhaps the discussion of this chapter rather implies), but is part of an already-successful theory. But, strictly speaking, my own feelings are irrelevant to my intentions, namely to give human sociobiology a chance to prove its scientific worth in the face of what I believe are so many essentially spurious criticisms.

What I am arguing, therefore, is that the most important matters at this point must be left primarily to the scientist. Eventually, human sociobiology will succeed or fail as a science. I am not a scientist; I am a philosopher, and for me, in John Locke's great words, "it is ambition enough to be employed as an under-labourer in clearing the ground a little, and removing some of the rubbish that lies in the way to knowledge . . ." (Locke, 1959, 1, p.14.) This I have tried to do and consequently my main task is now over. However, there is still perhaps some place for a philosopher. If human sociobiology does remain a viable research programme then it is going to expand and have implications, both within and without science. As a philosopher, one can legitimately and properly explore the ways in which one area of science can affect other areas, and one can also most certainly explore the non-scientific implications of a science. Therefore, concluding this book what I shall do is consider some of the possible implications of the success of human sociobiology: first in the realm of science, particularly the social sciences, and, secondly, in the realm of non-science, most specifically, given what some of the sociobiologists themselves have had to say, in the realm of the branch of philosophy having to do with the foundations of morality, namely ethics.

NOTES TO CHAPTER 7

[1] Objections have been raised to just about all kinds of study like this. For instance, it might be claimed that monozygotic twins tend to be treated differently from dizygotic twins. I shall be considering some pertinent objections.

[2] In fact, although I doubt the feasibility of direct studies on incest taboos by seeing if the children of incestuous unions are themselves incestuously inclined, later in the chapter I shall suggest that there is indeed some direct evidence bearing on incest taboos.

[3] Darwin did not accept a crude 'recapitulation' thesis: that, in ontogeny, animals exactly recapitulate their phylogeny. But he did believe, rightly, that there are some similarities between today's embryos and fossilized ancestors. (Ospovat, 1976.)

[4] On this matter, see Hull (1978a), for an entertaining and informative discussion.

CHAPTER 8

SOCIOBIOLOGY AND THE SOCIAL SCIENCES

To imploy a useful metaphor, up to this point we have been considering sociobiology from a 'static' viewpoint: that is to say, we have been taking sociobiology frozen at some point in time (the present!) and trying to evaluate its worth. In this chapter, I want to consider matters from a 'dynamic' viewpoint: that is to say, I want to see how things could develop through time, from the present to the future. In particular, assuming that human sociobiology remains a viable program and grows successfully — and note that in this chapter this is an assumption that I am making and because I have made it, it does not necessarily imply that it will prove true — I want to see what implications this might have for the rest of science. Since, obviously, the area of science which studies humans is (almost by definition) the social sciences, my inquiry is, in fact, an attempt to see how the growth of human sociobiology might affect the social sciences.

But, before I begin the detailed analysis, it will be worth making a few general points to guide us in our analysis. Referring to the history of science, let us ask what happens when one science, either a specific theory or a whole area, moves in on the domain of another science.

8.1. THEORY CHANGE: REPLACEMENT AND REDUCTION

Perhaps most obviously, when one part of science encroaches on another, the new part might push the old part aside, in the sense that the new part seems in important respects to be closer to the truth or more adequate in some way than the old part. Paradigmatic examples of this 'replacement' process are the Copernican revolution, where the heliocentric theory of the universe replaced the geocentric theory, and the chemical revolution, where Lavoisier's new chemistry replaced the old phlogiston theory.

In recent years, this process of replacement has come under intense philosophical scrutiny, for some philosophers have argued that replacement is a total matter, with everything that was held before, ideas, concepts, theories, being discarded and replaced by the new. Since the philosophers who have argued this way usually hold to some version of the thesis that the facts of science are theory-laden, that is that there is no pure observation but that all

165

sensing is done in terms of prior beliefs, it has usually been argued also that during replacement even the facts of science change: the world becomes a new place. (Kuhn, 1962; Feyerabend, 1970.) However, whilst there is much to be said for this position, many philosophers find it just too extreme. They cannot accept that at times of scientific replacement the world itself (or the world as we know it) actually changes: what they would argue rather is that the facts and even some of the interpretations remain. The change is in the overall theories. Thus, illustrating this more moderate view of replacement, take the Darwinian Revolution. In this revolution, a key, common fact was vertebrate homology; specifically, the skeletal isomorphisms between organisms of different species. The anatomist Richard Owen (1848) explained them in terms of a shared debt to the vertebrate archetype, a neo-Platonic transcendental blueprint on which all vertebrates were modelled. Charles Darwin (1859) explained them in terms of common ancestry, claiming that the vertebrate archetype was no ethereal reconstruction but a once-living organism. On this view, the world of bones has not changed; what has changed is the ultimate explanation. (See also Ruse, 1970, 1971; Hull, 1973.)

For the extremists, just about every change in science involves replacement. Hence, for them, science is a very discontinuous affair indeed! But those who oppose the extremists, feeling that even at times of replacement there is a continuity of fact, usually feel that there are many changes in science which are not even as drastic as their weaker form of replacement. The new science moves in on the old science, but it does not always push it aside: rather, in some instances, in some way, the new science absorbs the old into itself. Paradigmatic examples are the absorption of Galilean mechanics by Newtonian mechanics, and the macroscopic gas laws by the kinetic theory of gases. There is some debate about this process of 'reduction'; but the common view is that the older theory, the reduced theory, is, in some way, shown to be a consequence of the newer theory, where by 'consequence' in this context some sort of deductive link is implied. This means, in effect, that the older theory tends to be a special case or application of the newer more general theory which has overtaken it. (Nagel, 1961; Schaffner, 1967, 1969; Ruse, 1973, 1976. But see Hull, 1972, 1973, 1974; Darden and Maull, 1977; Kleiner, 1975.)

We see, therefore, that philosophers (not the same philosophers!) offer us somewhat of a spectrum of possibilities about scientific change: from an extreme all-out replacement, where even the facts change, *via* a weaker replacement which supposes a change of theories and interpretation, to reduction, where the older theory is taken up into the newer. To me, examples like both sides in the Darwinian revolution accepting vertebrate homologies make the extremist view of replacement implausible. Consequently, I suggest that in considering ways in which sociobiology might relate to the social sciences, we need only take the possibilities of the weaker form of replacement and of reduction. In order to help our analysis, let me make just one or two clarifying points about these.

First, although replacement (meaning, from now on, the weaker form) and reduction are usually presented as alternatives, I suspect that what often happens in science is that an older theory is corrected, which is a form of replacement, and then the corrected version can be reduced to the newer theory. Take, for example, a recent much-celebrated case of reduction, namely that of biological Mendelian genetics to physico-chemical molecular genetics. Understanding reduction to rest on deduction, in fact classical Mendelian genetics cannot be deduced from molecular genetics, if only because the Mendelian gene, the unit of inheritance and function, cannot be split apart: it is 'particulate'. The molecular gene however, the DNA, can be broken right down into small units, nucleotide pairs. A deduction is therefore impossible because the two theories are inconsistent. Nevertheless, if the Mendelian gene concept is revised, to allow division within the gene, consistency can be achieved and the possibility of reduction is reopened. And this is what has indeed happened, for the old biological gene concept has been replaced by the cistron, the unit of function, the muton, the unit of mutation, and the recon, the unit of crossing over. A cistron, containing many mutons and recons, can be identified with the molecular gene, and a recon with a nucleotide pair. It should be added that this correction was not simply a function of a desire to reduce biological genetics, but stemmed from problems internal to Mendelian genetics (e.g. the so-called 'position' effect) and from new biological methodology (e.g. the use of bacteria for fine-structure analysis).

The second point to be made is that a reduction (or reduction-cum-replacement) is often in principle rather than in fact, and scientists do not rush to flesh out the details of their reductions: indeed, they show supreme indifference to such a task. Certainly, this is the case in genetics, where one has a few bold strokes, but no detailed rigorous inferences. In other words, reduction is more a guideline: a philosophy, if you like. It gives scientists direction, new avenues to explore, and helps relate their various findings and make sense of their various claims. Thus, because of his reductionist viewpoint, Lewontin thought that his finding of molecular genetic variation was germane to the biological problem of whether or not there is genetic variation within species. That scientists should feel no urge to fill out all of the boring details of a reduction, rather than exploiting its implications, is only to be expected. (Schaffner, 1976; Ruse, 1976a; but see Hull, 1976.)

Finally, about reduction or reduction-cum-replacement, there is what one might call more of a psychological point. Reduction, at least at first, tends not to be a very happy process. The scientists moving in, those of the reducing science, are usually somewhat contemptuous of the scientists whose territory they are invading. Their own science is going well (by definition!), and often they are not very sensitive to what the reduced science sees as difficult problems. On the other hand, the scientists being invaded feel threatened and resentful. Thus, in the case of genetics, the scientists from physics and chemistry saw most biologists as woolly-minded vitalists: 'stamp collectors.' (Watson, 1968.) Conversely, the biologists saw the physicists and chemists as insensitive materialists. (Mayr, 1969.) Hence, there was personal animosity and lack of scientific communication. But this kind of breakdown, as indeed the case of genetics well illustrates, does not have to be either permanent or crippling. Lewontin's work (1974) clearly shows how biology today draws fruitfully on molecular studies, for matters of extreme biological interest and importance are being tackled with the help of physicochemical techniques and ideas. (Ruse, 1977a.) And more generally, far from biology vanishing, as the sociobiology controversy amply demonstrates, biology flourishes as never before. Thus, what I would suggest is that a reduction (or indeed a replacement for that matter) can be a liberating and creative experience for all concerned. It certainly does not necessarily spell the end of all that those in the reduced science have long worked for; indeed, it might give them a terrific impetus and powerful new tools.

So much then by way of preliminary. Let us now turn to biology and the social sciences. Following some recent suggestions by Wilson (1977a), I want to look at the possible effects of sociobiology on anthropology, psychology, economics, and sociology. I take them in this order.

8.2. THE REPLACEMENT OF ANTHROPOLOGY

The concern of anthropologists is generally with primitive or pre-literate groups of people, at least with groups which have not felt the full force of the Western industrial revolution (at which point those studying such groups turn into sociologists!) That is to say, anthropologists' concern is with peoples who are in some vague sense fairly close to nature. This would seem to imply, therefore, that if biology is going to have any real relevance to the human dimension, it is in the realm of anthropology that it might first surface. And indeed it does seem that anthropology already is the social science closest to sociobiology, although as we have seen from the reactions of the eminent anthropologist, Marshall Sahlins, this closeness is not something appreciated or even acknowledged by all anthropologists. (Our discussion in the introduction above leads us to expect such hostility, and confirms the closeness of sociobiology and anthropology — if the former were not moving in on the domain of the latter, practitioners of the latter would be feeling no threat.)

So, how might the scenario unfold? Less colloquially, how might the development and success of sociobiology affect anthropology? It would seem that there are various possibilities depending on the different positions the scientists involved can, and do, adopt. Let us start by considering fairly extreme positions, both of anthropologists and of biologists. These are positions which seem to point at best, assuming that sociobiology has some success, to some kind of replacement of anthropological theorizing by biological theorizing. At one extreme, there are anthropological positions which, essentially, deny the direct relevance of biology to the understanding of anthropological data. There are, I think, two forms of this kind of extreme position.

On the one hand, one might just deny outright any relevance at all of biology for anthropology. This, I take it, might be the stance of an extreme empiricist who wanted to eschew all anthropological theorizing and who thought his or her main task was describing the practices and beliefs of primitive societies. Whilst I doubt that anyone is ever as purely empiricist as he or she thinks he or she is, a name which springs to mind is Franz Boas (and his school) from earlier this century. Obviously, for a purely descriptive anthropology, the success of sociobiology is going to mean use of their findings and a replacement of biological theory for the theory that there can be no theory!

On the other hand, one might allow that biology has some relevance to anthropological theory, but not as a science which can or could, in its own right, explain that which anthropology explains. What I mean here is that one might deny that biology can itself explain anthropology (either theory or facts), but yet allow that biology can provide a fruitful source of models and metaphors to inspire anthropological theorizing. Obviously, one thinks here of people who hold to some kind of extra-biological, cultural evolutionary theory: in this century, Leslie White and his followers, prominent amongst whom is his student Marshall Sahlins.

Thus, we find White writing:

In the man-culture situation we may consider man, the biological factor, to be constant; culture the variable. There is an intimate generic relationship between man as a whole and culture as a whole. But no correlation, i.e., in the sense of a cause-and-effect relationship, can be established between particular peoples and particular cultures. This means that the biological factor of man is irrelevant to various problems of cultural interpretation such as diversities among cultures, and processes of culture change in general and the evolution of culture in particular. (White, 1959, p.418.)

And yet he and his followers have borrowed heavily from biology, arguing that cultures evolve in a law-bound fashion, just as organisms evolve biologically in a law-bound fashion. One should add that the borrowing has not always been of currently accepted biological theory (nor, since it was being used as a model, did it have to be). Sahlins and Service (1960) explicitly (and truly) deny that their main debt is to Darwin and the neo-Darwinians. They are much closer to someone like Herbert Spencer, for they see progress in cultural evolution and change from homogeneity to heterogeneity. Moreover, for them, the key to change is not survival and differential reproduction, but the harnessing of energy. "It seems to us that progress is the total transformation of energy involved in the creation and perpetuation of a cultural organization . . . The total energy so transformed from the free to the cultural state . . . may represent a culture's general standing, a measure of its achievement." (Sahlins and Service, 1960.)

Again, if sociobiology is going to have any meaningful impact, some replacement of this kind of anthropologizing is going to have to occur. Humans as biological beings are going to have to come into the scene, and not just accepted as background constants and then ignored. However, this said, perhaps total replacement may not necessarily occur. One may feel that human beings do introduce a new cultural dimension into the picture, perhaps even of a kind suggested by the above cultural evolutionists, and thus some of their ideas can be incorporated into a new overall theory. What I have in mind here obviously is some sort of theory in which a culture's ability to harness energy might be related to the effects that this would have on the survival and reproductive chances of individuals having such a culture. Although I shall not here be trying directly to incorporate a cultural evolutionary theory of the kind of Sahlins and Service into biology, in line with suggestions made in the last chapter, I shall shortly be looking at work which might be thought to take the first step in the direction of such an incorporation.

So far we have been considering the extreme anthropological position; that is to say, we have been considering the position(s) of anthropologists who really do not want anything to do with biology in its own right. There would seem to be a correspondingly extreme biological position: one which more-or-less denies the need for any distinctive anthropological theorizing about humans, arguing that anything which is of interest can be explained in precisely the same way that one would explain for any other biological entity.

Just as ants instinctively sense out what various relationships are, and do that which best perpetuates their own genes, so also, for all that they may think otherwise, do humans. Culture, as such, basically has nothing to do with the matter: it is just an epiphenomenon on top of the biology, perhaps serving only to conceal the reality from us. Obviously, if this position were to prove true, there would be just about total replacement of anything in anthropology at the theoretical level. Certainly, any attempt to give causes in terms of learning and so forth would be rejected, although presumably one would not reject the facts gathered by anthropologists.

I have suggested in the last chapter that, even though the way some of the sociobiologists write gives the impression that they believe that the future of sociobiology holds a replacement of anthropology of the kind just sketched, probably in their more reflective moments the sociobiologists would not argue in such an extreme way. Certainly, if sociobiology is to have any success at all, the extremist anthropological position — one denying any place at all for biology — will have to be replaced; but probably the sociobiologists do not see the replacement coming through a sociobiology which denies any place at all for anthropological theory. Rather, one suspects that many of the sociobiologists would not look unfavourably upon an extension of biology into the anthropological realm along the lines just hinted at in the last chapter.

This is to say, as biologists the sociobiologists would feel that one just has to start with the human individual as a biological entity: this means one must try ultimately to understand things in terms of the reproductive interests of the individual human. However, the sociobiologists would feel also that one cannot just dismiss culture and learning as unimportant: these are, in respects, the most distinctive human attributes of all. Hence, culture must in some sense be seen as adaptive for the individual human. No longer do humans have all their behaviour strictly encoded in their genes; rather their genes give rise to certain capacities or ranges of behaviour, and the specifics are filled in by discovery and learning. The great advantage of this being, of course, that new adaptive information can be passed directly from phenotype to phenotype, without having to go through the genotype. This 'Lamarckian' way of trans- mission ('Lamarckian' in the sense that now direct interactions on the pheno- type are important in evolution), has an overall adaptive power in the sense that it is far more rapid than traditional methods of evolutionary change.

Whether this approach will ultimately prove really fruitful is, of course, something only time can tell. But already some scientists are working along these lines, and it may perhaps prove instructive to consider briefly one such example, namely an analysis of primitive war by William Durham. When this has been done, because so far I have considered only anthropological positions of a kind which would essentially deny the importance of biology as a science in its own right, I shall turn to the work of anthropologists which might in certain respects seem more sympathetic to the kind of approach about to be illustrated.

8.3. PRIMITIVE WAR AS ANALYSED THROUGH A BIOLOGICAL-ANTHROPOLOGICAL COMPROMISE

Generally speaking, as illustrated in Figures 8.1 and 8.2, Durham (1976a) favours the attempt to integrate biology and culture along the lines just suggested, and he believes that we are already at the point where some positive moves in this direction can be taken. In particular, Durham (1976b) believes that there are two reasons why primitive war (that is war between primitive peoples) is an especially good test and confirmation of the kind of compromise position being advocated.

First, primitive war has proven to be something of a theoretical embarrassment to non-biological anthropology, "which has tended to believe that human societies are functionally integrated systems, well adapted to their environments". (Durham, 1976b, p.401.) Think of some of the embarrassing questions which come up. Why should some primitive peoples battle with

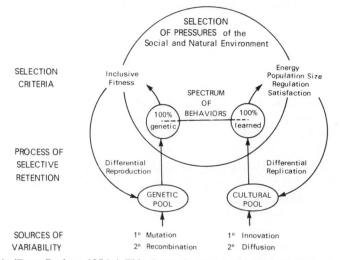

Fig. 8.1. (From Durham 1976a.) This shows the model of adaptive behaviours evolving through essentially separable biological cultural processes.

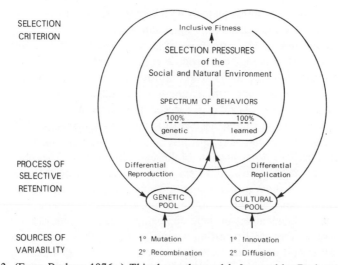

Fig. 8.2. (From Durham 1976a.) This shows the model, favoured by Durham himself, where the evolution of adaptive behaviour is a function of far more integrated biological and cultural processes.

each other, so fiercely, so continuously, and apparently so destructively? Take, for example, the Mundurucu, head-hunters of the Amazon. One anthropologist, who studied their history extensively, wrote:

... war was considered an essential and unquestioned part of their way of life, and foreign tribes were attacked because they were enemies by definition. This basic orientation emerged clearly from interviews with informants. Unless direct, specific questions were asked, the Mundurucu never assigned specific causes to particular wars. The necessity of ever having to defend their home territory was denied and provocation by other groups was not remembered as a cause of war in the Mundurucu tradition. It might be said that enemy tribes caused the Mundurucu to go to war simply by existing, and the word for enemy meant merely any group that was not Mundurucu. (*Ibid.*, p.404, quoting Murphy, 1957, pp.1025–6.)

But why should the Mundurucu be so aggressive? It seems no explanation at all to conclude, as did this anthropologist, that the head-hunters have a generalized aggressiveness and that war is a necessary "safety valve institution"; and it is at best *ad hoc* to tack on that old favourite, that such war preserves "the integration and solidarity of Mundurucu society". (*Ibid.*, p.404, quoting Murphy, 1957, p.1028.) Why do many other tribes, for example many Eskimo groups, not have such generalized aggressiveness? And why do the Mundurucu not find less costly ways of keeping their societies together? All of these questions ought to be answerable by an adequate anthropological theory, but they seem unanswerable given traditional approaches.

The second reason why primitive war is such a good test and confirmation of a biology—anthropology compromise is that, granting that traditional anthropology may have difficulty with primitive war, *prima facie* the integrated approach (*i.e.* the approach integrating biology and culture) seems no more promising. In other words, if it does succeed, then, as pointed out in a previous chapter, we have a confirming instance of the best kind, namely one of surprising or unexpected phenomena, not just something around which the theory was built. Consider: What possible biological value is there in chopping each others' heads off? Moreover, if we are going to take seriously the sociobiological claims about genes promoting an individual's reproductive self interests, what possible value is there to the individual in taking part in group war? Short of all of the members of the group being one's close relatives — unlikely indeed — one seems to be putting oneself at risk solely for the good of the group, and even if one's actions and causes are cultural, they are certainly not biologically adaptive in a sense accepted today. In other words, war seems just the sort of thing that a biologically influenced anthropology would exclude.

Nevertheless, argues Durham, there are conditions under which primitive wars (*i.e.* conflict between groups) could evolve and be maintained: conditions implied by and consistent with the principles of biology. What we need is a situation where there is a fairly close relationship between resources (particularly food) and the ability of an individual to survive and reproduce: that is to say, a situation where an increase in resources is going to have a fairly direct effect on reproductive chances. (See Figure 8.3.) Moreover, we need

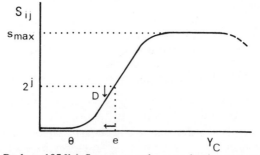

Fig. 8.3. (From Durham 1976b.) S_{ij} measures the reproductive success of an individual with respect to numbers of descendants j generations later. Y_c measures the consumption of a scarce resource. When $S_{ij} = 2^j$ we have 'equilibrium point'; that is, any less a value and the individual i is reproductively below par. Hence, one must consume at least e to have an equal representation of genes in the next generation.

a situation where just about everybody in the group is approximately in the same state with respect to resources: that is to say, everybody will gain and will gain about the same amount from an increase in resources – no one is going to help another person if the person helped gets virtually all of the benefits. (See Figure 8.4.) And, finally, one needs a situation where

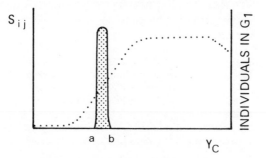

Fig. 8.4. (From Durham 1976b.) This shows the individuals in the group with respect to resources Y_C. Everyone falls between a and b, and hence everyone gets about the same amount of the resource (although over a number of generations the differences in resource attainment could have significant effects on reproductive success).

resources can most readily be increased by taking them from others or by preventing others from having them. In particular, with respect to this last point, one needs a situation where the resource demands of one's own group and of other groups overlap; where resources are just not sufficient to go round; where one is liable to come into contact with other groups; where alternative sources of resource increase (like agriculture) are not attractive; and so on.

If these conditions are not met, primitive war is not likely. Thus, the Eskimos

". . . cannot afford to waste time wrangling amongst themselves; the struggle to wring from Nature the necessities of life, that great problem of humanity, is there harder than anywhere else, and therefore this little people has agreed to carry it on without needless dissensions." Cooperation in the struggle for existence is absolutely imperative in their case . . . It is forced upon them by conditions of their environment (Durham, 1976b, p.392, quoting Davie, 1929, p.46.)

On the other hand, if these conditions just listed above are met, then it can pay (*i.e.* be reproductively advantageous for) an individual to join with her or his group in waging war on others. In the case of the Mundurucu, for example, the required conditions obtain. In particular, the Mundurucu are (and were) crucially dependent on animal protein, and, because of its relative scarcity, hunting for such protein is entirely time-consuming. Hence, wiping out competitors paid the individual, because then there would be more food for him and his kin. Consequently, war was of adaptive value to the individual.

Furthermore, in support of his hypothesis, Durham notes with some interest a phenomenon described but not appreciated by anthropologists, namely that in an important sense the Mundurucu identify the (human) enemy and game. After a war, with a village

... the most important status was that of a taker of a trophy head, who was referred to as *Dajeboisi*. Literally, the title means 'mother of the peccary', and allusion to the Mundurucu view of other tribes as being equivalent to game animals. The 'mother' part of the term is derived from the trophy head's power to attract game and to cause their numerical increase, and the head-hunter was so titled because of his obvious fertility promoting function: paradoxically for such a seemingly masculine status, he symbolically filled a female role (Durham, 1976b, p.405.)

If the integrative approach is correct, this phenomenon is easily explicable: the killer of the rival in an important sense raises the attainability of peccary, that is of game.

Finally, what about cheating? Durham invokes something akin to Trivers' reciprocal altruism. The individual that cheats and tries to avoid his or her share of danger will pretty soon be spotted and excluded from the benefits of warfare. And, in the case of the Mundurucu, Durham points out that there were strong social pressures which undoubtedly kept freeloading to a minimum. Everybody in the group had to play his or her role, whether it be going out hunting or staying at home and minding the camp.

We see therefore that, despite initial appearances to the contrary, a compromise biological—anthropological theory can give a plausible explanation of the evolution and maintenance of primitive war. It must be conceded that this analysis by Durham is but one example; but it does show the genuine possibility of building a theory which starts with humans as biological entities, subject to biological laws even in their social relations, and yet within these bounds pays full and open respect to the cultural dimension of humankind. This being so, let us now turn back to the work of anthropologists.

8.4. BIOLOGICALLY SYMPATHETIC ANTHROPOLOGY

When introducing anthropology in this chapter, I noted that already, in respects, it seems the discipline closest to sociobiology. Nevertheless, it might with reason be felt that up to now I have not given much evidence of this, for the only anthropological positions introduced so far (that is, the only positions of anthropologists introduced so far) have been ones which would call for a fair measure of replacement by a successful extension of sociobiology into anthropology. I would defend my approach, partly on the grounds that I wanted to show just what an anthropological extension of sociobiology would not allow, and partly because I think it is true to say that even today there is no one universally accepted way of doing anthropology and because undoubtedly there are still those (like Sahlins!) who would deny that anthropology properly done relates culture in any serious direct way to the biological concerns of survival and reproduction.

However, picking up my original point, it must in fairness be noted that there are now many anthropologists who are readily prepared to relate culture to biological ends; and thus in many respects they would not be totally unsympathetic to the kind of approach illustrated by Durham's analysis of primitive war, as given in the last section. In this kind of case, therefore, I think we might expect something more akin to the reduction of anthropology to biology, than its replacement; although for reasons to be given in a moment I am not sure that we will get quite the sort of reduction that we get in biology and I suspect that a certain amount of replacement will go on also. In particular, I suspect some rethinking will have to go on with respect to the question of who gains from adaptations, the individual or the group.

A good example of the kind of anthropological approach that I am alluding to at the moment is to be found in a recent textbook by Marvin Harris (1971). In theory, Harris is quite explicit about seeing culture (including learning, speech, and so forth) as adaptive in a biological sense. "Culture has made man the dominant life form of the earth." (Harris, 1971, p.38.) And, "... culture is man's primary mode of achieving reproductive success. Hence, particular socio-cultural systems are arrangements of patterned behaviour, thought, and feeling that contribute to the survival and reproduction of particular social groups". (*Ibid.*, p.141.) Moreover, in particular, Harris views culture as biologically adaptive. For instance, in a fascinating analysis of the sacredness of cows in India, Harris shows that although *prima facie* the Hindu taboos on killing cattle and eating beef are counter-biological, because the cows eat food needed by humans and do not themselves provide such food, in fact there are good biological reasons for making cows sacred. In particular, such cows provide dung, vital as fertilizer and as fuel. And, furthermore, since they are sacred and may not be eaten, the cows help Indians resist the temptation to turn from a vegetable-eating society over to a meat-eating society, something which requires a far more efficient agriculture than the Indians have. Eating up the cows, particularly in times of famine, could result in disaster. Finally, the cows do eventually provide lots of food for one segment of society, the pariahs who, being prepared to eat carrion, eat the cows when they die of old age or starvation or the like.

Clearly, this kind of approach taken to anthropology manifested by Harris' explanation of the cow taboo is not that far from the sort of thing the socio-biologists seem to envision. Although, as pointed out, some correction would be required, because Harris does not see culture as really being for the individual. "The relationship between the adoption of a socio-cultural innovation and its reproductive advantage is usually indirect and related to the reproductive advantage of the social group as a whole rather than to any particular set of individuals." (*Ibid.*, p.151.) Interestingly, in his discussion of primitive war, Harris sees it as being generally adaptive to the group because it keeps population numbers under control, but because he does not view matters from the biology of the individual, he sees it as a somewhat inefficient process which is at times maladaptive.

Of course, in saying an approach like this would synthesize fairly nicely with a sociobiological approach, I am not thereby implying that the anthropologists themselves would be happy to admit to this possibility of a synthesis, or that they would accept all of the particulars of sociobiological theorizing. Harris, for instance, deliberately rejects any explanations of incest taboos in terms of direct biological benefits, arguing instead somewhat along the lines of Levi-Strauss: namely that such taboos lead to, or are caused by, the exchange of women between groups controlled by men, something culturally valuable in that the process helps cement alliances and the transmission of information.[2] Parenthetically: I cannot myself understand why the Pharaohs marrying their sisters is taken to be such a devastating counter-example to biologically reinforced incest taboos. It is rather like arguing that the existence of albino negroes implies that the black skin of negroes cannot be genetic. And, as Alexander (1977b) points out, when dealing with people who have much to lose by marrying out of the nuclear family, the calculus of reproductive interest might imply rare breeding practices.

Enough evidence has now been presented, I think, to suggest that even if the anthropologists themselves might not entirely like the advance of socio-biology, in a sense the kind of work that many of them are doing is preparing the way. Finally, therefore, in this discussion of biology and anthropology, let us look a little more closely at the formal relationship between the two: if not as it is now, as it might be expected to develop as the two come even closer together.

8.5. THE FORMAL RELATIONSHIP BETWEEN A CORRECTED
ANTHROPOLOGY AND BIOLOGY

Let us suppose, perhaps somewhat condescendingly, that anthropology has been corrected to bring it in line with biology (*e.g.* by relating the ends of culture to the individual not the group), and that sociobiology continues to flourish. How might the two mesh together? Given the taxonomy at the beginning of this chapter, the most natural slot in which one might be inclined to put their relationship is 'reduction', since nothing very much is being replaced. However, in an important respect the situation envisioned differs from a reduction which seems to have occurred in genetics. There, we have a corrected Mendelian genetics (often known as 'transmission' genetics) being deduced from molecular genetics, or more precisely from molecular genetics together with all sorts of other assumptions like translation principles (allowing us to go from talk of strips of DNA to cistrons, mutons, recons, and so forth).

Now here, obviously, there is no question of deducing primitive war or the sacred cow taboo from the Hardy—Weinberg law. Not even with the aid of translation principles! Rather, what we seem to have is the biological theory being extended to the cultural context, but the biological theory being silent on certain key questions and therefore having to be supplemented by cultural anthropological principles. The situation is not that dissimilar to that obtaining where biogeography is derived from population genetics, together with a great number of principles peculiar to the disciplines. One cannot deduce the effects of trade-winds from the Hardy—Weinberg principle!

Let me try to show diagrammatically what I think is occurring. Normally, we think of the effect of selection as shown in Figure 8.5.

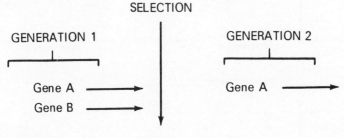

Fig. 8.5

(I am simplifying drastically and ignoring kin selection and the like.) Of course, this diagram is far too simple. In particular, we have forgotten all about the environment. At the very least we should have that shown in Figure 8.6.

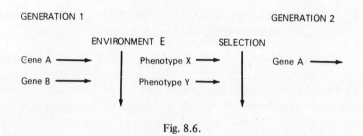

Fig. 8.6.

Even this is too simple. For instance, the two organisms might have different environments and this might make a crucial difference to the phenotypes' fitness. However, normally the environment is thought sufficiently stable not to be too major a causal factor as compared to the genes (at least, this is the assumption made). But, in the cultural case, what we are saying is that because of the kinds of genes involved, namely genes for knowledge (*i.e.* genes for thinking up new ideas, holding on to those ideas when useful and rejecting them otherwise, and for transmitting or learning ideas), the environment, or rather its differences, now become far more crucial. If we understand the environment as including the cultural mileau, what we now get is as shown in Figure 8.7.

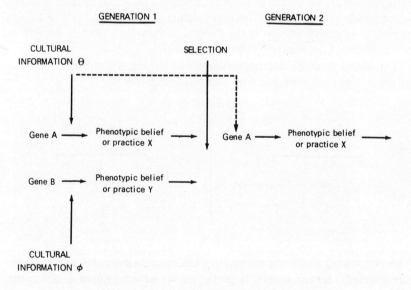

Fig. 8.7.

I have shown two different genes involved here. In the purely biological case, one suspects that matters would not be very interesting unless two different genes were involved. Nevertheless, in the case integrating biology and culture, the genes might be the same, although one suspects that, sometimes, different genes leading one to favour different kinds of beliefs or practices might be involved.

This is still obviously a very much simplified picture. For instance, it says nothing about new information, genetical or cultural. But the main point can now be grasped. Biological theory provides the outlines, leaving gaps which are then picked up by cultural theory: the nature of cultural information, how it affects the phenotype, how it is passed from one generation to the next, how it gets involved in selection, and so on. Anthropological theory is not deduced from biology; but it is integrated into it. Thus, we do not deduce the cow taboo from the genes. We do not even say that some Indians have (or had) genes which made them crave beef and others have genes which make them abhor beef, and that the latter were selected over the former. What we do say is that those with the cultural belief which makes them abhor beef are selected over those without such a belief. This is not to deny that there might be genes (or combinations of genes) which would make people more readily accept taboos of some kind, and that these sorts of genes have been accepted. Unless people could have fears and phobias, beef taboos would get nowhere. This point obviously gets us close to a discussion of genes influencing certain kinds of learning, a topic which we shall be taking up in the next sections.

Anthropology, therefore, can help to fill in the biological gaps; and, in return, biology can provide an overall theoretical framework for anthropology. As pointed out in the case of genetics, that the exact relationship may never be spelt out in a rigorously formal way is not too worrisome (and unless something goes wrong, is hardly to be expected). It is the outline and possibility that counts. But, assuming things do develop or show the potential of developing in this way, should we speak of the biological-anthropological relationship as 'reduction', or not? Probably, it is mainly a question of semantics and it certainly does not matter very much. The situation being hypothesized is not quite the same as occurred in something like genetics; but it does involve a development of theory without replacement (or, at least, no more replacement than in the genetics' case) and demands no radical revisions of our views of theories. If, for instance, one thinks the hypothetico-deductive ideal is in some way appropriate, what is going on here seems to fit in nicely.[3] The important point is that, perhaps even more dramatically than in the genetics' case, we see that the arrival of biology on the anthropological scene does not threaten anthropology. We do not even have anthropology being deduced from biology. It is filling in areas indicated by biology but untouched by biology. In this sense, we have something which is not ominous for, and destructive to, anthropology; but exciting, liberating, and extending.

8.6. PSYCHOLOGY: THE PROBLEM OF LEARNING

Psychology is of particular interest to our discussion, because whatever the exact formal situation may be or prove to be, psychology illustrates nicely a phenomenon mentioned in the case of genetics: namely that when one science (or theory) moves towards another, the second science may be moving towards the former for reasons internal to itself. Thus, although molecular biology moved towards Mendelian genetics, Mendelian genetics also moved towards molecular biology as it was corrected into transmission genetics. Something not too far different seems to have happened between biology and psychology, particularly over a phenomenon of great interest to both: learning. Not that long ago, biologists and psychologists were far, far apart on this question. But now they are much closer, and even though sociobiology might hope to take matters even further, this coming together was not just the work of biologists, but stemmed from psychology also. Let us look at things a little more closely.

On the one hand, one had the biologists, particularly the so-called 'ethologists', prominent amongst whom were Konrad Lorenz and Niko Tinbergen. (See especially Tinbergen, 1951.) For them, it was a basic premise that learning must be understood in an evolutionary context. This means obviously that different situations call for different responses.

The student of innate behaviour, accustomed to studying a number of different species and the entire behaviour pattern, is repeatedly confronted with the fact that an animal may learn some things much more readily than others. That is to say, some parts of the pattern, some reactions, may be changed by learning while others seem to be so rigidly fixed that no learning is possible. In other words, there seem to be more or less strictly localized 'dispositions to learn'. Different species are predisposed to learn different parts of the pattern. So far as we know, these differences between species have adaptive significance. (Tinbergen, 1951, in Seligman and Hager, 1972, p.245.)

Thus, for example, the herring gull very quickly learns to tell its own chick from that of another; however, despite significant differences it cannot tell its own eggs from that of another. But then, chicks have a tendency to run away; eggs do not. On the other hand, although a gull cannot tell its own eggs, it learns rigidly the site of its nest. It will even ignore its own eggs if they are taken out of the nest, despite the fact that the eggs may be placed in full view of the brooding bird. Again, the adaptive significance of such learning is obvious. And much the same tale is told through the animal kingdom; perhaps the best-known phenomenon being 'imprinting', when, at a crucial period in its life, some animals learn in an irrevocable way to identify with and only with those animals in whose company they find themselves.

Most psychologists, the so-called 'general-process learning theorists', either ignored or denied these kinds of views. They followed Thorndike, who had written: "If my analysis is true, the evolution of behaviour is a rather simple matter. Formally the crab, fish, turtle, dog, cat, monkey and baby have very similar intellects and characters. All are systems of connections subject to change by the laws of exercise and effect." (Bitterman, 1965, in Seligman and Hager, 1972, p.310.) And, not unexpectedly, this kind of position tended to be self-reinforcing. Because there were thought to be no major differences between organisms, research tended to concentrate on one organism, namely the rat; and then because all the research was on one organism, there was no counter-evidence to the belief that all organisms learn in the same way!

Furthermore, not only was biology denied or ignored in the claim that learning varies not at all from species to species, it was thought irrelevant to what actually is learned. One starts with a conditioned stimulus, say eating some food, one applies an unconditioned stimulus, say shock, and one gets a response, say future aversion to that food. And this was thought quite general. "... any natural phenomenon chosen at will may be converted into a conditional stimulus ... any visual stimulus, any desired sound, any odor, and the stimulation of any part of the skin." (Seligman and Hager, 1972, p.3.) In a very real sense, the biology of the organism does not count at all.

Obviously, if biology, traditional ethology or sociobiology, was to make any progress against a psychology such as this, a fairly substantial replacement was going to be required. Moreover, the facts cited by psychologists had to be shown to be a biased sample of the whole. It is just inconceivable to a biologist, even if one were not to accept every claim of Lorenz and Tinbergen, that all animals should learn exactly all things in exactly the same way. It would be like all animals being exactly identical, not only between each other, but also between the various parts of their own bodies! However, in fact it has not proven necessary for biologists to move in on psychology, drastically replacing defective sections. From within psychology itself has come a reaction — or perhaps one should say a 'revolt', for showing how strong was the feeling that biology is irrelevant, it might be pointed out that only 10 years ago *Science* refused to accept work which went against the dominant psychological paradigm (the 'indifference hypothesis'). (Seligman and Hager, 1972, p.8.)

The barriers have been breeched by the pioneering work of John Garcia and his associates, and following in his footsteps there is an ever-growing recognition by psychologists that success in learning is a crucial function of the kind of learning involved and of the species of organism involved. (Garcia and Koelling, 1966; Garcia, Ervin, and Koelling, 1966; Garcia, McGowan, and Green, 1972.) In particular, Garcia exposed rats to a taste and to an audio-visual stimulus, followed by radiation sickness one hour later. Contrary to the indifference hypothesis, only the taste became aversive. On the other hand, when given footshock rather than radiation sickness, it was the audio-visual stimulus rather than the taste which became aversive. Findings like these, which have since been reduplicated in various forms many times, obviously go against general process learning theory. The rats ought have reacted equally to taste and to audio-visual stimulus, and further-more an aversion ought not have built up given so long a gap (one hour) between conditioned stimulus (taste) and unconditioned stimulus (radiation sickness).

But, as has been readily recognized by many psychologists, these findings of Garcia make perfect sense within a biological evolutionary framework. Food is liable to poison one, whereas audio-visual stimuli are not. More-over, a symptom of food-poisoning is a general sickness like radiation sick-ness. Hence, there is a selective premium on learning to avoid foods which make one feel sick, and clearly one is that much fitter if one can become averse to a food even though the sickness does not become immediately apparent, because this is the way that many poisons work. In short, be-cause of the genes that they have — genes which have been selected because of their adaptive value — rats become averse to tastes when followed, even when followed some time later, by sickness. Because there has been no similar selection for genes causing audio-visual stimulus aversion (linked with sickness), rats do not become averse to audio-visual stimuli in the same situation.

It has become increasingly clear that this example of the rats is by no means unique and that much animal learning is selective and makes best sense when considered from an evolutionary perspective. Thus, to take but one further rather nice example, indigo buntings orient their autumnal migrations according to the position of the heavens. But in a series of elegant experi-ments using the Cornell planetarium, Stephen Emlen (1970) was able to show that what the buntings learn is not some fixed map but the axis of rotation of the heavens. Exposed to heavens rotating around fictitional axes, the birds 'migrate' according to them. This plasticity in the birds' learning has an obvious evolutionary value, for, to take an instance, in 13,000 years the North Star will no longer be Polaris in the Little Dipper but Vega in Lyra, 47 degrees away. In the evolutionary time-scale 13,000 years is but little time, and, were the buntings to have a fixed orientation to the heavens, before long they would be migrating to all sorts of unsuitable places. As with the rats, natural selection has been at work here.

Moreover, it is clear that human learning is also selective in a way that points to biological causes. For instance, many of us have had experiences like that of the rats. Martin Seligman vividly describes an aversion to Sauce Bernaise, which was brought about by stomach flu following on a meal including such a sauce. (Seligman and Hager, 1972, p.8.) Why, asks Seligman, did he not develop an aversion to the china plates from which he ate his meal, to the opera which he saw that evening, or even to his wife who shared the meal? Clearly selection has primed his genes to avoid food followed by stomach pain, or more precisely, selection had primed his genes to avoid *strange* food followed by stomach pain, for he developed no aversion to the steak beneath the sauce! Similarly, when it comes to phobias, we tend to show our evolutionary history. For instance, most of us shy away from snakes and the like, something of obvious adaptive value. On the other hand, to our frequent downfall we tend not to react the same way to knives and electric outlets and other dangerous things around the house. Here, as elsewhere, we show the effects of our biology, for we learn rapidly to avoid things which in the wild would be highly dangerous (and for which therefore there would be strong selective pressure for genes making us ready to avoid them). Conversely, something like an electric outlet, unknown in the wild, leaves us unmoved. (Seligman, 1971.)

It would seem therefore that, unlike ten years ago, with respect to learning there is a growing recognition by psychologists that one ignores biology only at one's peril. Animal learning, including human learning, has been fashioned by natural selection. Although much work obviously yet remains to be done, for instance no one yet seems to have much idea about which genes might be involved, it does seem that we are here pointing to a reduction of psychology to biology. I take it that here, as in the case of anthropology, rather than a deduction of human psychology from anthropology, what we are going to have is psychology working in areas indicated by biology, so that our understanding of the genes and of the learning environment come together in a unified whole.

8.7. PSYCHOANALYTIC THEORY AND THE EXPLANATION OF HOMOSEXUALITY

What about parts of psychology, other than that dealing directly with problems of learning? One area which springs immediately to mind, given the interests of the sociobiologists themselves, is psychoanalytic theory. And indeed, Wilson writes: "Psychoanalytic theory appears to be exceptionally compatible with sociobiological theory ... If the essence of the Freudian revolution was that it gave structure to the unconscious, the logical role of sociobiology is to reconstruct the evolutionary history of that structure." (Wilson, 1977a, p.21.) Rather than attempting an overview of psychoanalytic theory and its possible relationship to sociobiology, let us just look for a moment at one phenomenon of great interest to people on both sides: homosexuality.

One thing upon which all seem to agree is that there is most probably no one cause for homosexuality, or even, restricting ourselves yet further, for male homosexuality. But, as we have seen, gathering together their explanations, the sociobiologists want to lay the causes of homosexuality in various ways at the door of the genes: that there are genes for homosexuality because of balanced superior heterozygote fitness; that there are such genes because of kin selection; and that there are genes for parental manipulation bringing about such homosexual behaviour. Now, of course one cannot speak for all psychoanalytic theorists, but the impression one gets is that they certainly would not rule out the possibility of a genetic factor in some homosexuality. At least, in their discussions the heritable possibility is raised and the evidence presented. (For instance, Mamor, 1965.) By any standard, the evidence is not overwhelming and certainly there are definite cases that do not fit it. But most psychoanalytic theories (*i.e.* those in print!) seem not totally opposed to the idea that the genes might play some small role in homosexuality. In this case, presumably, one has a fairly direct explanation of psychological facts by biology.

However, it must be fully admitted that for psychoanalytic theorists the environment, particularly the home environment during development, is the crucial causal factor in one's becoming homosexual rather than heterosexual. And, moreover, whilst not everyone is a total Freudian by any means, there seems general agreement that Freud was right in suggesting that behind much homosexuality lies a parental imbalance. In particular, as is well known, Freud suggested that male homosexuality comes about when the mother is far more dominant than normal, running both the father and the children. (Freud,

1906.) Psychoanalytic theorists seem prepared to go this far with Freud, although it is perhaps another matter as to how many would go all of the way with him, suggesting that the male child is in love with his mother, that in normal development the incest taboo causes him to transfer his sexual interest to other females, and that when the mother is very dominant the ties are too strong to break and thus the child's oedipal guilt causes him to recoil from women in general and to transfer his emotions to men.

Prima facie we seem therefore to have conflict between sociobiologists and psychoanalytic theorists about most of the causes of homosexuality. But it is clear that the conflict is not necessarily so very great at all, for at least one of the sociobiological mechanisms, that centering on parental manipulation (Trivers, 1974), seems almost tailor-made for the psychoanalytic theorizing. Remember, in this mechanism the pertinent genes are possessed by the parent, especially the mother, not the child who turns into a homosexual. The hypothesis is that the mother manipulates one of her children into being a non-reproducing homosexual (where this manipulation is certainly not necessarily a conscious process). It takes no great leap of the imagination to suggest that the manipulation takes the form of the mother's becoming over-protective of a younger son, for we know that there is statistical evidence that homosexuals are more likely to be younger than older sons. (Pare, 1965.) In short, one can easily suppose that the way that the genes bring about the required behaviour is by setting up precisely those conditions which the psychoanalysts hypothesize is required for bringing about homosexuality! In other words, the psychoanalytic ideas fit nicely into the gap which is pointed to by biology. The two theories, biological and psychological, do not conflict, They complement each other.

As in the case of anthropology, I am not quite sure that it is true to say that psychoanalytic theory has been reduced to sociobiology, as biological genetics has been reduced to molecular genetics. Giving a diagram similar to that given previously, the situation seems to be somewhat as shown in Figure 8.8. At least, this is the situation from the biological viewpoint, which just leaves blank the move from parental manipulation to homosexual behaviour in the offspring. This is the point at which psychoanalytic theory can move in. Thus, perhaps the best thing to say is that we have one area or theory of science incorporated within another. If we want to think of this as a kind of reduction we can; but it is not a formal deduction. The main point, of course, is that biology calls for psychoanalytic theory, which latter can flourish as never before.

Please note in concluding that, on the basis of the homosexuality example,

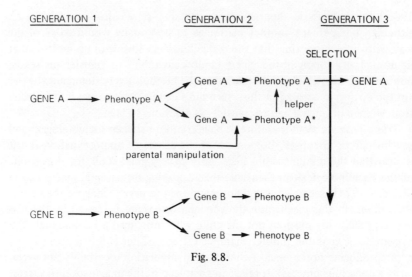

Fig. 8.8.

I am not suggesting that the sociobiologist demands or endorses every psychoanalytic theoretical claim ever made. One certainly does not have to be an orthodox Freudian to be a sociobiologist. One might, for instance, want to put some other causal explanation than the Freudian one between the dominant mother and the homosexual son. Sociobiology in itself would not stop one. The point is that, with respect to homosexuality, sociobiology positively beckons a psychoanalytic explanation.

Generalizing this cautionary note about causes, it should perhaps be added that sociobiology would not necessarily be tolerant of absolutely every psychoanalytic suggestion, as it is of the analysis of homosexuality. Although both sociobiologists and Freudians think the incest taboo of fundamental importance to humans, I am not sure that the sociobiologists could tolerate Freud's (1913) story of its origin in the successful conspiracy of a band of brothers to kill and eat their father (although no doubt this tale of family strife would strike a responsive chord in Trivers breast!) But then, I am not sure how many psychoanalytic theorists today would be able to tolerate this story, taken totally literally.

Obviously, therefore, a certain amount of replacement of psychoanalytic theories, either by the sociobiologists or the theorists themselves, would be required to bring sociobiology and psychoanalytic theory into harmony. But one feels that it probably could be done; that given the joint interest in so many topics, it might be worthwhile that it be done; and that certainly such a project does not mean the triumph of sociobiology and the end of psycho-analytic theory. Homosexuality requires such a theory. Incest taboos, even if backed by the genes, require an explanation of why brother and sister set up aversions to breeding. And the same is true time and again elsewhere. Thus, as in the case of anthropology, the coming of sociobiology to psychology spells hope and renewed vigour for the social sciences; not their demise.

8.8. ECONOMICS

A personal value judgment is that we have now covered those areas of social science which show most promise for a successful and fruitful integration with sociobiology.[4] Apart from anything else, we have covered those areas of social science where their subject phenomena are least heavily covered with a thick layer of culture, that is to say where we are closest to the underlying biological phenomena. I am not suggesting that, for example, the anthropologist studying primitive tribes does not encounter culture — of course, such a person does, and indeed my discussion starts from that fact. But as explained earlier, it is in such areas as anthropology that we might expect the biology most clearly to show through. And, in fact, if there is anything to what has so far been written in this chapter, even if all other areas are excluded, the meeting between biology and social science will be most fruitful for all.

Of other areas of social science, economics and sociology spring most readily to mind when one starts asking questions about the future of sociobiology. Taking, first, economics, although I have just admitted to the suspicion that we have covered those parts of social science most ready for sociobiology, I do not want to imply that there is no place or possibility for fruitful interaction between biology and economics. In fact, there seem to be at least two avenues which are worth exploring. First, there is the possibility of revisions to classic economic theory. Second, there is the possibility of a comparative economics. Let us explore these two avenues briefly. (See also Wilson, 1977a.)

In classic economic theory, humans are taken as rational economic machines. For instance, to take a recent textbook chosen quite at random, the author begins with three assumptions about consumer preferences which are, in his words, "quite plausible". (Mansfield, 1970, p.22.) First, a consumer can always decide which of two options he or she prefers, or whether each option is equally valuable: whether he or she prefers one basketball ticket and 3 chocolate bars, or 4 bottles of soda and a bus ticket. Second, consumer preferences are transitive. If one prefers blonds to brunettes and brunettes to redheads, then one must prefer blonds to redheads. (His example, not mine!) Third, one always prefers more of a commodity rather than less. "For example, if one market basket (a very big one) contains 15 harmonicas and 2 gallons of gasoline, whereas another market basket (also big) contains 15 harmonicas and 1 gallon of gasoline, we assume that the first market basket, which unambiguously contains more commodities, is preferred." (*Ibid.*, p.22.) Moreover, not only are humans taken as rational (if these assumptions are rational!), but it is assumed that one can properly theorize at the level of the group rather than the individual, and this group is taken to be a rational entity. For example, the same textbook introduces the notion of a firm, "a unit that produces a good or service for sale". (*Ibid.*, p.114.) And, furthermore, this firm is assumed to work as an integrated whole to maximize its own profits.

Clearly, all of these assumptions and ones like them are highly suspect. Human beings just do not always act in the way that classical economic theory supposes. Often humans are 'irrational' in their choices, and certainly they do not always act as an integrated group (in fairness to the author of the above-mentioned textbook I must note that he recognizes this fact in an appendix). In recent years, a number of economists have been facing up more and more to the realities of the human situation, and trying to reflect such realities in their idealized models. For instance, recently, Harvey Leibenstein (1976) has put forward a theory based on the notions that central to economic theory must be the individual not the group, and that we must pay full attention to something which he calls "x-efficiency", this being a function of various psychological factors like motivation, enjoyment of task, and the like.

Leibenstein argues that rather than looking upon individuals as beings who try to maximize their economic gain, we must look upon individuals as beings who are always trying to compromise between the things that they would like to do and the things that one ought to do. "In order to describe the characteristics that govern economic behaviour it helps to distinguish among capabilities, the *desire* to use capabilities under constrained circumstances, and the desired standards that individuals want to meet. In general, individuals have to compromise between two sets of opposing psychological forces: the desire to use one's capacities outside the bonds of the constraints inherent in a context; and the desire to fulfill the demands of one's superego, that is, the desire to meet as much as possible one's internalized standards, which in part depend on the observed performance of others." (*Ibid.*, p.93.)

I am not endorsing Leibenstein's theory, and even if one did we are still a long way from biology. But it is clear that there is a need for social psychology in economics, and that, in fact, economists are not insensitive to this need and that some indeed are trying to do something about the matter. Perhaps, therefore, if and when psychology can be brought closer to biology, this may in turn bring economics and biology closer together. As we try to understand what makes human beings work, why they do not behave as classical economists would like, we may well find ourselves referring to human evolution.

The second possible avenue of exploration for interaction between biology and economics is that of an interspecific comparative economics. As we know, sociobiologists draw heavily on economic theory for their models. By his own admission, Wilson's recent work on the division of labour in insect societies (written in collaboration with George Oster) is so far dependent on the social science that it rather resembles a text-book in microeconomics. (Oster and Wilson, 1978.) There are, admittedly, aspects of the insect situation which differ from the human world: for instance, insect societies are mostly sterile and female and the altruism shown is far greater than in the human case (because of the haplodiploid genetic peculiarities of Hymenoptera). But, for all of the differences, Wilson feels emboldened to draw some pretty strong conclusions. "The point is that human economics is not really general economics, but rather the description of economic behaviour in one mammalian species with a limited range of the biological state variables." (Wilson, 1977a, p.23.)

I must confess that I am not sure that this conclusion is justified, at least as yet. It is one thing to use one area of science as a model for another area; it is another thing to suggest that this modelling shows that the two areas are fundamentally part and parcel of the same thing. Apart from anything else, we have obviously got a significant cultural element in the human world. We clearly have got to work out in some detail the relationship between culture and the genes, before we can say to what extent, if any, human economic behaviour relates to the genes or can be treated as if it related to the genes. And then and only then can we start to enquire meaningfully into the question as to the extent that human economic behaviour is really the same as the economic behaviour of other organisms which latter behaviour is undoubtedly essentially a function of the genes. There may prove to be a substantial overlap. As we have just seen, Leibenstein argues that we must consider the individual as primary in economics, not the group, and I do not now have to tell the reader that this is a major move in the direction of general sociobiological thought. But we are a long way yet from showing real identity and not just a useful heuristic analogy. In other words, I am not ruling out the possibility of a comparative economics, but I cannot see it on the near horizon.

8.9. SOCIOLOGY

Finally, we come to sociology: the study of human social behaviour, particularly in more advanced (*i.e.* more industrialized) societies. As Wilson (1977a) admits, the hope of biology having any influence here seems at present dim indeed. In fact, it is bad enough trying to introduce any psychology, for many sociologists share Durkheim's deeply anti-reductionistic bent. "In a word, there is between psychology and sociology the same break in continuity as between biology and the physico-chemical sciences. Consequently, every time that a social phenomenon is directly explained by a psychological phenomenon, we may be sure that the explanation is false." (Wilson, 1977a, quoting Durkheim, 1938.) Few, perhaps, would be quite this blatant, but *de facto* the extent to which sociology draws on psychology is slight indeed. And indeed, although there are certainly areas of exception, sociology tends not to be that theoretical at all. Certainly, as compared to something like biology even, let alone physics and chemistry, sociology comes across as a low level descriptive science.

Durkheim has proven gloriously wrong about the relationship between biology and physico-chemical science. To me, it is inconceivable that he will not prove wrong about the future relationship between psychology and sociology. The whole history of science points to the profitability of trying to relate one area of science to the area just above it, meaning by this latter the area which deals with entities at a slightly smaller conceptual level. And, obviously, if indeed sociology is brought closer to psychology, then inasmuch as psychology is in turn coming closer to biology, biology may perhaps make its influence felt at the sociological level.

But all of this, if it lies anywhere at all, lies in the future. In any case, I am certainly not arguing that sociology as we know it at the moment is just simply going to be deduced from the movements of the genes. I have not argued this for anthropology, and as we come to sociology the strictures and qualifications made there increase tenfold. In anthropology, we deal with societies which are long-lasting and close to nature; in sociology, we deal with societies which change rapidly and which give individuals far greater insulation from the brute forces. As Wilson says about Western cultures: "Having been jerry built on the Pleistocene human biogram, they are the least stable, probably have the greatest discrepancies between genetic and cultural fitness, and hence are most likely to display emergent properties not predictable from a knowledge of individual psychology alone." (Wilson, 1977a, p.25.) Here, probably, the contributions of sociobiology are going to remain slight, at least for some time to come.

8.10. CONCLUSION

The great German philosopher Hegel once deduced from logical principles that there can be only eight planets. Shortly thereafter a ninth planet was discovered. As a fellow philosopher, I am therefore reluctant to commit myself too heavily to predictions about the future course of science. This chapter, I must admit, has been speculative, despite the confident (too-confident?) tone of my assertions. But one thought which I took into the discussion, a thought which has been amply reinforced and which it seems worth re-emphasizing in conclusion, concerns the overall effect of the possible encroachment of sociobiology on the social sciences. Whether we have replacement or reduction, and whatever the kind of reduction involved, the coming of biology does not mean the end of the social sciences. As has happened for biology since the coming of physics and chemistry, it means the very opposite: there is hope for new techniques and ideas to tackle problems insoluble today, and for the opening up of whole new vistas of exciting research areas, as yet but vaguely discerned. Social scientists should welcome sociobiology, hoping that it will be successful: not recoil from it with fear and disgust.

NOTES TO CHAPTER 8

[1] I discovered Durham's work, which I shall be using as a paradigm of such an approach, through a favourable reference in Wilson (1977a).

[2] Even here, of course, we have an explanation in terms of indirect biological value.

[3] Undoubtedly, if one does not much like the hypothetico-deductive ideal, what is occurring can be made to fit with one's analysis of theories, to show that there is continuity rather than a break.

[4] If we consider the learning of a language separately from learning in general, then linguistics is obviously also an area where one might expect to see some fruitful interaction between sociobiology and the social sciences. Indeed, the approach of Noam Chomsky seems almost as if it were designed with sociobiology in mind. (See Wilson 1975a and 1977a for a brief discussion and references.)

CHAPTER 9

SOCIOBIOLOGY AND ETHICS

As soon as the *Origin of Species* appeared, indeed from before that date, there were scientists who wanted to argue that the only true moral philosophy is one firmly grounded on evolutionary theorizing. One thinks here specifically of Charles Darwin's contemporary, Herbert Spencer, who more than anyone drew tight bonds between evolution and ethics; although, given Spencer's present reputation, whilst philosophers think of Spencer as primarily a biologist, no doubt biologists think of Spencer as primarily a philosopher. (Perhaps the quickest way to achieve compromise is for both biologists and philosophers to agree that, essentially, Spencer was a father of social science.) However, despite the fact that Spencer's rather mushy ideas were countered with devastating rhetoric and logic by T. H. Huxley (1893), Darwin's 'bulldog' and in his own right an evolutionist at least as eminent as Spencer, various brands of 'evolutionary ethics' have kept appearing during the past hundred years. Amongst recent efforts along these lines, perhaps the best-known attempts have come from the pens of (of all people) T. H. Huxley's grandson, the late Sir Julian Huxley (1947), and that fascinating biological maverick, the late C. H. Waddington (1960) (in calling him a 'maverick' I mean no disrespect, rather the opposite).

These more recent attempts at evolutionary ethicizing have, in turn, been cut down, in the opinion of most people (*i.e.* most philosophers!) as effectively as T. H. Huxley cut down Spencer. (Flew, 1967; Quinton, 1966; Raphael, 1958.) Nevertheless, undaunted, the sociobiologists have decided to run a tilt or two in the lists. They feel that now, and only now, do we have the biology of humans clearly within our sights, but that since we do it is appropriate and proper to use our findings to explore at once all aspects of the human predicament, including the ethical. Indeed, so strong is this feeling by some sociobiologists that Wilson goes so far as to begin *Sociobiology* as follows:

Camus said that the only serious philosophical question is suicide. That is wrong even in the strict sense intended. The biologist, who is concerned with questions of physiology and evolutionary history, realizes that self-knowledge is constrained and shaped by the emotional control centers in the hypothalamus and limbic system of the brain. These centers flood our consciousness with all the emotions – hate, love, guilt, fear, and others

194

– that are consulted by ethical philosophers who wish to intuit the standards of good and evil. What, we are then compelled to ask, made the hypothalamus and limbic system? They evolved by natural selection. That simple biological statement must be pursued to explain ethics and ethical philosophers, if not epistemology and epistemologists, at all depths. (Wilson, 1975a, p.3.)

Taken literally, people like me ought to be put out of their jobs! (A sentiment, no doubt, upon which the sociobiologists and their critics will for once find agreement.) And even at his more conciliatory, Wilson rather thinks that we philosophers ought to take lengthy, enforced sabbaticals. "Scientists and humanists should consider together the possibility that the time has come for ethics to be removed temporarily from the hands of the philosophers and biologicized." (Wilson, 1975a, p.563.)

As might be expected, like everyone else we philosophers have genes for self-preservation (Socrates not withstanding), and no doubt matters will not seem quite so clear-cut to us. At least, stimulating and interesting though I find Wilson's suggestions I do not find them overwhelmingly persuasive, either his suggestions about philosophy or about philosophers! Perhaps indeed: "Philosophers and humanists should consider together the possibility that the time has come for biology to be removed temporarily from the hands of the biologists and philosophized". But enough of such tempting dreams. Let us, in this chapter, conclude this book by considering the possible relationship between evolution and ethics, paying particular attention to the work and the suggestions of the sociobiologists.

There seem to be at least three significant ways in which evolution and ethics might interact. (Munson, 1971.) First, evolutionary biology might throw light on the fact that we humans are ethical animals at all. Second, evolutionary biology might supply the theoretical foundation or justification for ethics. Third, ethics might help us direct evolution in the future. These three considerations are by no means entirely distinct, but for ease of exposition let us try to treat them as such and take them in turn. I should add that whether these considerations are 'truly' philosophical or 'truly' biological does not really interest me: for all of my desire to tease them for their presumptions, I share with the sociobiologists the conviction that these matters ought to be of concern both to philosophers and to biologists.

9.1. WHY ARE WE ETHICAL?

I trust that the reader will not think that in my sub-title I beg the all-important question: namely, are we ethical or moral at all? I think it will be generally agreed, except possibly by school-teachers and parents at the end of a long day, that all human beings show care and concern for others, even to the point of their own inconvenience or loss. (Whether one can be moral towards oneself is a nice philosophical point and need not concern us here.) Furthermore, it will be generally agreed — although perhaps not quite so generally agreed — that, at least at the phenomenological level, this concern for others is voluntary. We choose to put ourselves out and to help others. Understanding such attitudes and behaviour to be 'moral' or 'ethical', it cannot be denied that none of us are completely moral all of the time, and some of us are not very moral most of the time; but, for all of our faults, moral concern does seem to be a feature of humanity. And, interestingly, even the great fiends of history like Hitler have tended to try to cloak their misdeeds in a mantle of (spurious) morality.

Now, granting this moral sense or faculty that we have, and granting also that it is not something supernaturally imposed upon us, the matter of explaining its origin becomes pressing. Moreover, if one is a Darwinian evolutionist, that is if one believes that natural selection was a prime cause of evolution, the matter of explanation becomes even more pressing, for at least in some sense there seems a premium on self-interest: those organisms whose genes did not promote phenotypes which would do better than others in the struggle to survive and reproduce would not be those organisms passing on the most genes to the future. In other words, it would seem *prima facie* that morality does not pay from an evolutionary perspective, and hence ought not to have evolved.

As we know full well by now, there are various ways out of this dilemma. First, we might invoke some sort of group selection hypothesis, arguing that since morality (almost by definition) works for the good of the group, its causes must be a function of selection working at the level of the group. But, as we also know, there are serious scientific objections to this kind of hypothesis, and so having already earlier discussed these objections at length, we can drop this hypothesis at once without further argument. Second, it might be suggested that the human moral sense evolved without any adaptive function: it is just a side-effect of the rest of human functions. This explanation also seems less than satisfactory. Generally speaking, it is always a last resort appeal to suggest that things have no function; and more particularly to this case, it seems unlikely that so pervasive and significant an element of human nature would be just a side-effect, especially if one cannot show of what it is a side-effect. Indeed, since, on the surface, morality seems not in an individual's evolutionary interest, one might have expected it to have been selected against. Hence, one needs a strong argument to show why this should not be so. No one, I take it, wants at present to suggest that the genes for moral sense are pleiotropically linked to the genes for eyesight!

Third, one might suggest that the moral sense and all of its consequences are entirely cultural: that biological evolutionary theory has no relevance at all to the origins of morality and ethics. However, whilst it is undoubtedly the case that many particular aspects of ethical beliefs and practices are a direct function of culture, to argue that everything of this kind is caused through cultural development seems rather to miss the point. Although there are in contingent fact aspects of culture which are not to do with morality, in an important sense morality seems a necessary condition for human culture (and, conversely, morality may well be a sufficient condition for culture, although it would indeed be a stripped down version of culture with morality alone). In other words, what I am suggesting is that culture could not itself be a cause of morality as such (although it may mould particular aspects of morality) because culture in some sense presupposes morality – unless people can work together in a giving and sharing manner there is no culture. Therefore, it would seem that in some important sense we must look for a biological base for morality, or rather for the human moral sense. And, in any case, we have already seen reason to believe that human culture in some overall sense must be biologically adaptive, and because moral behaviour is such a large part of culture, it too despite any appearances to the contrary must surely be adaptive. Otherwise, we have to admit that even in the most primitive of peoples, a large portion of their behaviour is probably seriously dysfunctional.

Fourth, therefore, almost by a process of elimination, we must turn to the sociobiologists, seeking explanations for morality in terms of selective advantages for the individual. And, as we know, the sociobiologists are happy to oblige, explaining the evolution of the human moral sense in terms of such mechanisms as kin selection and reciprocal altruism. I do not think that even the most enthusiastic sociobiologist would want to argue that the whole task of explaining the evolution of morality is completed; but it would be claimed that the essential outline is now sketched. Morality, or more particularly the moral sense, comes about because the moral human has more chance of surviving and reproducing than the immoral person. The immoral person fails to help relatives and does not get help from non-relatives, because they, in turn, can expect no help from him or her.

I do not want to raise again the already over-discussed question of the truth of human sociobiology. For the sake of argument, assuming in this chapter what we assumed in the last, namely that sociobiology is basically sound, I think one would have to agree that, with respect to the question of the causes of morality, sociobiology represents a significant step forward. The alternatives are less than adequate; but more importantly, the evolution of morality flows naturally from basic premises of sociobiological theory. Hence, in this sense, in answer to our question, "why are we ethical?", sociobiology seems a vital move forward. Note, however, that I am talking about the *causes* of morality, or rather of one's moral sense. As we shall see full well in the next section, I am not conceding that this is all that there is to ethics and morality: in particular, I am not conceding that ethics and morality are now 'justified'. Now, to repeat a point made earlier, am I conceding that humans are essentially 'selfish'. To talk of selfish genes is to talk metaphorically, and the whole point is that the phenotypes they promote are anything but selfish. A saint may be a product of evolution, but this is not to deny that he or she is a genuinely good person.

One final comment should be made in this section, which qualifies, but by no means obliterates, the reservations just drawn. Even though finding the causes of moral behaviour may not be to justify that behaviour, a fuller understanding of the causes of morality will surely have fairly direct implications for the bounds which we draw to moral behaviour. To be moral implies a choice: a falling stone can be neither moral nor immoral, despite the good or (more likely!) harm it does. People are moral because they choose to do the right thing and immoral because they choose to do the wrong thing. Now matters start to get complicated because it is clear that not all human actions are voluntary, that is involve choice. Sometimes, we are at the mercy of forces beyond our control: perhaps falling in love is something of this nature. And sometimes some of us cannot help doing that over which others of us have control. Moreover, even when we have choice, there are times when some of us are just not capable of telling right from wrong.

All of these facts lead us to say that under certain circumstances and at certain times some people are not morally responsible for their actions – we do not condemn them as we would a normal person. That is why in law we have such verdicts as "Not guilty by reason of insanity". Of course, the reasons why we should make such exceptions do not necessarily rest on genetic causes. Often, it is felt that something in someone's environment exonerates them. But clearly the reasons can and do sometimes rest on genetic factors. In other words, there are times when, because of a person's genes, we are less inclined to say that they are responsible for their actions, and hence less prepared to judge them morally. And it is not difficult to see that as our knowledge of the causes of human action is broadened through the development of sociobiology, the limits which we set on responsible human action may well be altered. We may well decide that certain things which today we fault (or indeed praise!) ought not be laid at the feet of the perpetrators. That is, the development of human sociobiology may alter the accepted domain of human responsibility for moral action. (For more on this kind of point, see Hudson, 1970.)

9.2. EVOLUTIONARY ETHICS

We come now to the controversial area of possible interaction between evolutionary theory and ethics. Even if we concede, as we have conceded, that evolutionary theory can throw light on how ethics and the moral sense came about, how they were caused, is it the case that evolutionary theory can in some way *justify* ethics, or tell us how we ought morally to behave? It has certainly been thought in the past that it can, and the sociobiologists are, in significant respects, of the same way of thinking. Let us begin with a few general remarks referring to the past, and then turn to see what Wilson has to say about these things.

The claim of the traditional evolutionary ethicist is that the course of evolution shows us what is good. In other words, what has evolved is good, and our moral obligation must be to further and aid the workings of nature as revealed through evolutionary theory. Of course, as might well be imagined, matters are not quite as simple as all this. Different evolutionists have had different ideas of what evolution is all about and how consequently one might best further its course. Thus, entirely contradictory moral norms have been drawn. Spencer, for instance, saw evolution as a kind of progression, from 'homogeneity' to 'heterogeneity'. This meant, in fact, that it was a kind of progression up through the monkeys, *via* the lowest forms of human life like Tierra del Fuegans and Irishmen, to the highest forms which were (honesty compelled Spencer to confess) something very much like middle-class Englishmen. And in order to turn us all into fine specimens of *Homo britannicus*, Spencer thought we ought to give the struggle for existence a free hand, adopting a *laissez faire* economics and social system, ruthlessly letting the weakest in our society perish. (Spencer, 1850, 1852a, 1852b, 1857.) However, others, for example Prince Kropotkin, believed in some form of group selection, so they saw evolution as involving altruism towards members of one's own species, and thus they advocated entirely opposite actions from those of Spencer towards one's fellow humans! (Himmelfarb, 1968; Peel, 1971.)

Now, how are we to judge arguments like these, whether we side with

Spencer or with his opponents? Until recently, it has been generally agreed that they all fall before an argument formulated by Hume but made popular in this century by G. E. Moore (1903). In particular, they make the mistake of going from 'is' to 'ought'; they assume incorrectly that one can legitimately infer "This is how things ought to be" from "This is how things are". Hume writes:

In every system of morality which I have hitherto met with I have always remarked that the author proceeds for some time in the ordinary way of reasoning, and establishes the being of a God, or makes observations concerning human affairs; when of a sudden I am surprised to find, that instead of the usual copulations of propositions, *is* and *is not*, I meet with no proposition that is not connected with an *ought* or *ought not*. This change is imperceptible; but is, however, of the last consequence. For as this *ought* or *ought not* expresses some new relation or affirmation, it is necessary that it should be observed and explained; and at the same time a reason should be given, for what seems altogether inconceivable, how this new relation can be a deduction from others, which are entirely different from it. (Hume, 1740, quoted in Flew, 1967, p.38.)

In this century, the logical flaw which Hume identifies has come to be known as the 'naturalistic fallacy', a label applied by Moore who thought that those who go from 'is' to 'ought' commit the mistake of identifying a property of one kind with a property of another kind. In particular, when one argues (as for example the utilitarians did) that something like happiness is the supreme good, the thing which one ought to strive to maximize, one is identifying a non-natural property with a natural property: for Moore, happiness was a natural property which we sense, like blue or warm, whereas good is a non-natural property which we do not sense but intuit.

Now, it is fairly clear that the evolutionary ethicist commits the naturalistic fallacy. One is going from "This is the way that the world *is* (because of evolution)" to "This is the way that the world *ought* to be (and hence help evolution to keep up the good work)". We go from (for instance in Spencer's case) "Human beings evolved through natural selection" to "We ought to let natural selection continue unhindered".

Unfortunately, we cannot leave matters simply here, arguing that evolutionary ethics commit the naturalistic fallacy and that is that. (Quinton, 1966.) Just as many biologists have recently changed their minds about the workings of selection, so many philosophers have recently changed their minds about the naturalistic fallacy. They have now decided that it is no fallacy at all, and just as scientists prefer their Great Men to be without blemish, so there is no shortage of voices prepared to claim that Hume meant precisely the opposite to that which for years everyone had been taking him to mean!

Whether one accepts these counter-arguments or rejects them (as I am inclined to do), it must be admitted that they have a *prima facie* plausibility. Thus, to take one popular example of a supposedly legitimate bridging of the is—ought dichotomy, in reply to the question "which hotel should I stay at?", the (ought) statement "You ought to stay at the Red Lion Hotel" seems clearly implied by the (is) statement "You will most enjoy staying at the Red Lion Hotel." Hence, rather than getting into a lengthy philosophical digression, it will perhaps be better if evolutionary ethics can be countered other than by simple invocation of the supposed naturalistic fallacy. (Warnock, 1967; Hudson, 1970.)

But even without the fallacy, traditional evolutionary ethics seems unsatisfactory. I take it that if someone is proposing an ethical theory then they must put forward some reasons as to why we should accept it: for example, that it accords with common decency, and that perhaps it systematizes and makes more explicit our beliefs. It may indeed call for us to revise some of our present beliefs and habits — for instance, it might ask us to become vegetarians — but it can do so only on the grounds of such things as consistency with other beliefs we hold dear. One might of course stipulate a new ethical theory (*e.g.* one ought to be nice to people under 5' 6" and horrid to those over); but without reasons, every one else is at perfect liberty to ignore it.

Now, I assume that an evolutionary ethicist would want to say that the course of evolution is, in fact, that which we would want to call 'good'. It has certainly been so in the past. Perhaps whether it will necessarily be so in the future depends on whether one thinks that evolution as a contingent matter of fact is good or whether one thinks that unfettered evolution is necessarily the good; but in either case common prudence would surely dictate that one ought to try to give the forces of evolution a free hand. (Of course, if one defines whatever happens as 'evolutionary' then one can do what one likes, but then one has no guide for moral action at all.)

However, in the light of what has just been said about evaluating ethical proposals, suggestions for basing ethics on evolution seem just not sensible, nor do suggestions that ethics is based on evolution seem true. I do not necessarily want to speak for everybody's commonsense views on morality, but I expect most feel as I do, namely that there is considerable truth in both the proposals of Kant and of the utilitarians (Ewing, 1953): on the one hand one ought to treat human beings as ends not as means, and on the other hand one ought to try to maximize happiness (whether, if pursued fully, these are overlapping, distinct, or contradictory views, need not concern us here — the whole point about common sense is that it tends not to dig too deeply! But, for all that, it is a good guide).

The trouble is that, whether one be a Kantian, a utilitarian, or both, an evolution-inspired ethics clashes with one's moral views. Take, for example, the smallpox virus. This is a product of evolution: one which the World Health Organization is trying to eliminate. But, inasmuch as WHO is trying to eliminate smallpox, it is trying to frustrate the course of evolution. It is trying artificially to make one species extinct. Yet surely, no one would want to say that the actions of the members of the WHO are morally wrong, and that people (including ourselves!) should just be allowed to die from small-pox, hoping that as in the case of myxomatosis with the rabbits, selection will see that the virus grows less virulent and we grow more resistant. Eliminating smallpox shows a concern for people as ends and an increasing of happiness — hence, on both counts we call it 'good'. Thus, it cannot be that the course of unfettered evolution is a good thing, or that we ought to promote it.

The obvious move for the evolutionary ethicist at this point is to argue that the primary concern is not with evolution *per se*, but with the evolution of humans. It will be said that we ought promote the forces that led to and keep up the evolution of humans. This being so, it can be seen easily that the elimination of smallpox, because it is a threat to humans, is a good thing not a bad thing. Unfortunately, this move will not work either. First, it presup-poses part of the very thing one is trying to establish, namely that evolution is a basis for ethics. One is importing things very like a Kantian or utilitarian ethic, humans as ends or human happiness, in order to establish one's main claim, that the prime good is the evolution of humans. How else can one justify such a restriction? Second, one is not, in fact, ruling out the smallpox example. Since undoubtedly the existence of smallpox affects human evolu-tion (*i.e.* succumbing to smallpox is in part a genetic phenomenon and thus there is selection for smallpox resistance), one ought rather to look upon smallpox as a good thing!

Third, more generally considering past human evolution and its present effects as goods in themselves goes against our intuitions. As Wilson points out, happiness (which we would consider a good) and the state of being adap-tive by no means necessarily go together. Speaking of human aggression and how unhappy it makes us all at times, Wilson concludes:

The lesson for man is that personal happiness has very little to do with all this. It is possible to be unhappy and very adaptive. If we wish to reduce our own aggressive behaviour, and lower our catecholamine and corticosteroid titers to levels that make us all happier, we should design our population densities and social systems in such a way as to make aggression inappropriate in most conceivable daily circumstances and, hence, less adaptive. (Wilson, 1975a, p.255.)

Fourth, it is by no means the case that letting the future course or human evolution proceed unfettered is a good thing. We are today caught in a population explosion, which already causes a lot of unhappiness. If it continues unchecked, then it is going to cause a great deal more unhappiness, because a lot of people are going to die through disease, war, famine, and the like. And the fact that those that survive, if indeed some do, have genes somewhat different from those that do not survive, is not going to minimize the unhappiness or constitute a good in any other way. This fact incidentally has not escaped the sociobiologists who have argued that we ought morally do something about it. Indeed, Wilson argues that we ought to act quickly and strive to achieve "healthier and freer societies". (Wilson, 1975b, p.50.)

Of course, Wilson and everyone else seem to think that we ought to keep the human species going; but whilst generally I would agree with these sentiments, so far does the promotion of the human species seem to me to be from the ultimate good, that I can imagine circumstances where one might want to argue that the human species ought to be brought to an end. Suppose there were good evidence that the earth was going to enter a radio-active zone 100 years hence, and that nothing we could do would prevent all humans then alive dying slow and painful deaths. I suggest that we would now have a moral obligation to end all breeding: both on Kantian and utilitarian grounds. In other words, there are principles of ethics more basic than the promotion of the human species and its evolution.

The consequence of these arguments is that traditional evolutionary ethics will not work (and indeed, even Spencer was not above also invoking an appeal to "a surplus of agreeable feeling" as well as to evolutionary progress). However, the sociobiologists seem to feel that the case is not yet closed. This is, on the face of matters, a little strange. We have just seen how some of the facts and the arguments of the sociobiologists themselves can be used effectively to counter traditional evolutionary ethics. Moreover, Wilson knows of, and apparently endorses, the standard moves against evolutionary ethicizing!

The moment has arrived to stress that there is a dangerous trap in sociobiology, one which can be avoided only by constant vigilance. The trap is the naturalistic fallacy of ethics, which uncritically concludes that what is, should be. The 'what is' in human nature is to a large extent the heritage of a Pleistocene hunter-gatherer existence. When any genetic bias is demonstrated, it cannot be used to justify a continuing practice in present and future societies. (Wilson, 1975b, p.50.)

And yet, as we have seen, Wilson now believes that the time has come for philosophy to be 'biologicized'. How can this be and what does this entail?

9.3. WILSON'S ATTACK ON INTUITIONISM

There are two parts to Wilson's assault on the problems of ethics, the combined effect of which is an argument concluding that we must accept evolution, its results and its processes, as the good, because in some sense almost by definition that is what the good has to be. At least, I think this is what Wilson's conclusion is, although as I shall show, he is contradictory. The two parts to Wilson's analysis are first an attack on what he takes to be the accepted justification of ethical positions and secondly an affirmation of moral relativism based on the findings of sociobiological theory. Let us take them in turn and see where they lead us.

First Wilson attacks what he sees as the primary philosophical justification offered today for ethical views, namely *intuitionism*, that is "the belief that the mind has a direct awareness of true right and wrong that it can formalize by logic and translate into rules of social action". (Wilson, 1975a, p.562.) He is not entirely clear as to what he finds wrong with this position, but the primary fault seems to be that it does not take into account the fact that the organ of intuition is a product of evolution.

The Achilles heel of the intuitionist position is that it relies on the emotive judgment of the brain as though that organ must be treated as a black box. While few will disagree that justice as fairness is an ideal state for disembodied spirits, the conception is in no way explanatory or predictive with reference to human beings. Consequently, it does not consider the ultimate ecological or genetic consequences of the rigorous prosecution of its conclusions. Perhaps explanation and prediction will not be needed for the millennium. But this is unlikely – the human genotype and the ecosystem in which it evolved were fashioned out of extreme unfairness. In either case the full exploration of the neural machinery of ethical judgment is desirable and already in progress. (*Ibid.*, p.562.)

One feels still one would like to have spelt out a little more clearly the reason why this should all count against intuitionism, but, reading a little between the lines, presumably the main cause for complaint is that because the brain is a product of evolution, we cannot rely on its perceptions or judgements or what have you in a way that is centrally presupposed by intuitionism. We know that different people have different evolutionary interests. We know, from sociobiology, that people will 'see' that which it is in their (evolutionary) interests to see. This seeing or perceiving is not necessarily of the truth. Hence, an ethical intuitionism, that is an ethical belief that we have a direct insight into the moral truth, whether it be that we ought to maximize happiness or that we ought to treat humans as ends, just cannot be accepted. It is all too possible (nay probable) that our genes are deceiving us and filling us full of a glow of having achieved absolute truth. Totally deceived moralists are far more evolutionarily efficient than conscious hypocrites.

I think this is Wilson's position. Certainly it sits well with the moral relativism that we shall see him adopting shortly. And certainly it seems the kind of position that Trivers would support, for he is quite explicit in his belief that because of evolution we cannot trust — ought indeed mistrust — the perceived or intuited 'truths' we hold dear: ". . . the conventional view that natural selection favours nervous systems which produce ever more accurate images of the world must be a very naïve view of mental evolution." (Trivers, 1966, p.vi.)

Before turning to examine this argument critically, in fairness to philosophers as a tribe it should perhaps be noted that it is a little strange that they should be saddled with the claim that intuition is the main support of ethical judgements. It is true that in the first part of this century intuitionism was popular, but for nearly fifty years now — indeed from the rise of logical positivism — other meta-ethical theories have found favour with many philosophers. (Hudson, 1970.) One thinks of emotivism, its descendent prescriptivism, and, more recently, naturalism. Indeed, without necessarily endorsing any one of these positions or, in fact, rejecting intuitionism, on the surface it would seem that intuitionism was a rather unfair choice to show that evolution destroys philosophical justifications of ethics. Actually, another choice might have led entirely to the opposite conclusion, for something like emotivism seems almost tailor-made for the evolutionist. The emotivist, in fact, entirely side-steps the difficulties that the sociobiologists think that evolutionary theory raises for the ethicist. When the emotivist says that one ought to do x, what he or she thinks one is saying is that he or she approves of doing x and "Do thou likewise". The truth-claim refers to one's own personal feelings, and (without wanting to get unduly psychoanalytic) seems to be an introspection the truth of which not even sociobiology can take away from one. And, for the emotivist, the rest of a moral claim is exhortation, which is neither true nor false. In other words, ethical statements for the emotivist cannot involve the possible divorce from reality, which the sociobiologists seem to think can make so suspect philosophical claims about ethics. (Ayer, 1946; Stevenson, 1944.)

But this is all a bit by way of preliminary: By making his case against ethics in terms of intuitionism, Wilson rather strikes me as being like a philosopher who rejects genetics because he or she finds fault with the classical gene

concept of T. H. Morgan. However, even against intuitionism the case is not
as devastating as all that. At least, I think one can invoke a strong and effective *tu quoque* argument. Every argument which can be made against ethics,
can be made against other truth-claiming statements, particularly those of
science — and even more particularly those of sociobiology! In other words,
using sociobiology to undercut ethics is hopelessly circular. Consider: The
supposed problem with ethics is that we get at it only through evolved organs,
and unfortunately these might lead us astray because it may well be in our
evolutionary interests to be deceived. But, with respect, how do we get to
know the facts of science, or of mathematics, or of logic, other than through
organs that have evolved through natural selection? Of course one might argue
that these organs would not deceive one, but that surely is to assume the
whole point! If they are deceiving us, then because we use these very organs to
understand them, they will fill us with confidence about their veracity.

   Nor can it be argued that biology shows that deception would come only
over matters of morals and not over matters of science and logic. It is clear
that our science and our logic are just as much of adaptive value as is our
ethics, and so deception is possible. Moreover, if one argues that a mark of
possible divorce from reality is a chopping and changing of minds (not possible
if one is tuned right into the truth), and since morality seems so changeable
this shows ethics not absolutely true, I would suggest that science seems no
less changeable — for all that the science of the day seems so overwhelmingly
certain. Compared to the two-thousand-year Christian code, astronomy seems
positively fickle. One can still follow Socrates in ethics; one would look a bit
silly following Ptolemy in astronomy. Finally, if it be argued that ethics cannot be intuited because different people arrive at different conclusions —
children and idiots, for instance, have trouble with understanding morality —
exactly the same argument can be brought against science. My small children,
for example, certainly have at least as much of a grasp of the differences
between right and wrong as they do of the principles of modern physics.

   In short, the case against intuitionism does not succeed. One might, of
course, conclude that what the above argument shows is not that intuition is
infallible but that all of our knowledge is fallible, in which case, presumably,
one has to adopt some sort of pragmatic attitude, arguing that one assumes
what works for the time being. But even here, ethics is no worse off than anything else, and one certainly cannot use science against it. Simply speaking,
what has gone wrong with the sociobiological argument at this point is that a
confusion has been made between *causes* and *reasons*. (Raphael, 1958.) It is
more than likely that we have ethics and a moral sense because of evolution,

that is through evolutionary causes. This does not mean that the reasons, the justifications of ethics, are evolutionary, any more than the fact that we have science and mathematics because of evolved organs means that the reasons for the principles of science and mathematics are evolutionary.

9.4. WILSON'S MORAL RELATIVISM

We come now to the second part of Wilson's argument. Having supposedly disposed of philosophers' justifications of ethics, he argues that different people have different evolutionary interests and thus we are stuck with a total moral relativism. Sociobiology shows that different people, young and old, female and male, have different evolutionary interests. But:

If there is any truth to this theory of innate moral pluralism, the requirement for an evolutionary approach to ethics is self-evident. It should also be clear that no single set of moral standards can be applied to all human populations, let alone all sex-age classes within each population. To impose a uniform code is therefore to create complex, intractable moral dilemmas — these, of course, are the current condition of mankind. (Wilson, 1975a, p.564.)

Fortunately, this appalling conclusion is no more well-taken than Wilson's previous conclusion about intuitionism. For a start, given his previous argument, Wilson has absolutely no right to talk of 'moral pluralism', or moral anything else for that matter. If, as I understand him to think he has done, he has negated philosophical (or other rational) justifications of ethics, then all we are left with are organisms with different, clashing, evolutionary strategies. The only difference between the human case and any animal case is that humans lay over their strategies this layer of beliefs that there are genuine moral standards. But, essentially, that is, from the viewpoint of the logic of morals, humans are no different from the animals: there is no 'real' morality.

However, if this is so, then Wilson ought not talk about moral pluralism. A pluralism of desires maybe; but when did a desire automatically have moral force? I want the chocolate cake. My sister wants the chocolate cake. There is no question of morality here. We certainly do not need to invoke a theory of moral pluralism. The only way in which we could achieve such a conclusion in Wilson's case is if we were to argue that, traditional arguments for morality having collapsed, we are therefore entitled to define morality as that which is in the interests of an evolutionary strategy. But this is at best a stipulative definition, and should not be presented as the outcome of an analysis of how we do and ought properly to use the word 'moral'.

The second point about Wilson's argument is that there is a conflation between different levels, of a kind noted more than once already in this book. We, or more precisely although metaphorically, our genes, have different evolutionary strategies. But as we know full well, at the phenotypic level, which gets us to the level of actual desires, of culture generally and of moral beliefs particularly, we do not necessarily have a plurality of wants (which Wilson's argument presupposes). Indeed, what we find is that although people have different evolutionary strategies, even manifesting different desires, they tend to share the same moral code. In other words, in another sense they want the same thing.

Consider: I expect most heterosexual men have felt as I do occasionally that there is some particular woman that they find sexually attractive and that they would really like to go to bed with. Now, from a biological view-point, this is all to my interest. If I can carry out my desires than I may well impregnate yet another female and thus pass on yet more genes (assuming that the foetus is not aborted, and so forth). And yet, without wanting to appear unduly burdened with a moral sense, I think I can genuinely say that because of my moral beliefs there are times when I have even stronger wants not to have intercourse with a woman toward whom I feel a strong sexual urge: suppose, for example, that she is a married woman and that if her in-fidelity were discovered it would cause great hurt to her whole family, includ-ing herself. The conclusion from such a case seems to be that although we may all have different evolutionary strategies, we may all want the same moral code — even those of us who break it! Hence, even our desires do not necessarily commit us to a moral pluralism.

Finally, in case it is objected that people do not all share the same moral code and that the differences may well represent different genetic back-grounds as the results of different evolutionary forces, I would suggest that this problem may well be overcome by drawing a distinction between differ-ent levels of a moral code. Suppose it is pointed out that, in the West, we practice monogamy and that this is backed by moral sanctions (at least, it used to be!); but that some societies practice polyandry, several husbands with one wife, where this is considered ethically acceptable if not obligatory. Suppose also that this is offered as evidence for moral relativism, and that this were backed by pointing to Alexander's (1974) explanation of polyandry in terms of parental manipulation. One can still argue — indeed, I would argue — that all of these different marital practices can be fitted under a higher com-mon rubric, namely that all people ought to have a chance to get married. And this rubric, in turn, fits under a totally universal Kantian or utilitarian ethic: that people not just be means (*e.g.* sex objects for others' gratification) and have the opportunity for maximum happiness, they ought to have the opportunity of long-term relationships with a partner (this is not to imply that this ought to be obligatory or that it is what everyone would want). Hence, because different situations have different needs, all does not collapse into a morass of relativism.

Wilson's arguments therefore do not work. This is perhaps just as well, for he himself blithely ignores his conclusions as soon as he has drawn them, arguing quite inconsistently that our present existential predicament demands that we start planning for the good of the whole, putting aside selfishness. Of course, what one might argue, and what I myself have indeed argued, is that although human culture as it presently stands is both a product of evolution and, in some general sense, biologically adaptive, it gives us the power to transcend our biology in certain respects. This means that we are no longer helpless pawns of our biology, but can act morally even though our basic desires drive us in other ways. Thus, suppose we accept (what I do not accept) that biologically, in some absolute way, men are dominant over women, this is not to deny that we have reached the point where our culture enables us to alter this state: that by freeing women from childbearing and so forth, we can now direct women into equal power with men. However, even if one argues this way, or rather because one can argue this way, Wilson's pluralistic selfish moral relativism fails. And the same holds if one tries related tactics, arguing for example as Alexander (1971) does sometimes, that we humans have now got to the point where our selfish interests and group interests coincide. There is nothing to stop us from consistently accepting human sociobiology and rejecting moral relativism.

9.5. CAN EVOLUTION BE DIRECTED?

Confining our discussion strictly to the level of human evolution which is a direct function of changes of gene ratio, although this restriction will be lifted later, the third way in which ethics and evolution might interact is if in some fashion we can gain a measure of control over evolution. Then obviously we must invoke ethical principles in discussing how evolution ought to be directed. From our viewpoint therefore, the specific questions of importance are whether evolution, particularly human evolution, can be directed, and if indeed this is so whether sociobiology throws any light on the way that it ought to be directed?

As far as human evolution is concerned, it cannot be denied that humans have been responsible for altering its course. If we think of the forces of evolution as consisting essentially of natural selection working on random mutation, both sides of this equation have been altered by humankind. Natural selection has obviously been directly tampered with inasmuch as we now save people (who then become reproducers) who, because of genetic ailments, would not otherwise have survived and reproduced. One thinks for example of people with various kinds of diabetes, ailments known to have a genetic cause. Today, they can live full, active, and reproductive lives, thanks to insulin. However, this means that they are now passing on their defective genes, whereas previously the genes would have died with them. Hence, in this sense we are altering the course of human evolution, because we are shielding people from the effects of natural selection. Probably in other senses, more indirect, we are again altering the way selection affects evolution. Indubitably, modern life increases stresses and strains, and this may well set up selection against certain genes. For instance, O-blood group types (caused by a certain gene) seem more prone to ulcers, caused by stress, than do other blood group types (*i.e.* other gene carriers). Of course, it might, with truth, be pointed out that today, even people with ulcers tend to live long enough to breed, so how great this particular kind of selection is may well be questioned. But, the general point holds, and certainly in the past humans have intensified natural selection. The move to urban living brought on conditions for the spread of T.B., and it seems likely that there is a genetic factor in susceptibility to the disease; hence, there was selection against certain gene types. And similarly, the missionaries not only took Christianity to the poor, benighted heathen, but white people's diseases like influenza: whites had genetic immunity against the full effects of these diseases, but the unprotected savages were killed off like flies. (Dobzhansky, 1962; Ruse, 1974.)

Turning to the other factor in evolution, mutation, humans have altered this too. Thanks to the publicity that has been brought to nuclear weapons and their fall-out, little more need be said on this subject here. One suspects that atom bombs are not the only culprits, however. Given the number of food additives that lead to cancer, it would be odd indeed were none of them mutagens.

The conclusion, therefore, is that humans have altered and do still alter the course of their evolution. And this is despite the fact that many people today have the odd illusion that, thanks to modern technology, human biological evolution has ground to a halt. If anything, we have speeded evolution up! But, it might with truth be objected that alteration is not really direction. By letting off atom bombs we are hardly 'directing' the course of evolution. However, in recent years, the power of direction has started slowly to come within our grasp. Science fiction delights in stories of selective breeding or of cloning or of genetic manipulation, but the first of these methods of directing evolution seems morally somewhat repellent (apart from being dreadfully slow) and although great progress is being made on the others, for humans we have some way to go yet. (Assuming that recent unsubstantiated reports of a cloned human are fictitious.) Nevertheless, we are getting more skilled at locating and detecting genetic causes of illness, and at finding these causes early in human development: one thinks here of the technique of amniocentesis, whereby amniotic fluid can be drawn from a pregnant woman and hence cells of the unborn child can be obtained and studied. With the growing preparedness of people to permit abortions, this means that many carriers of genetic diseases can be destroyed before they are born, let alone before they can, in turn, reproduce and spread on their defective genes. (Hilton, 1973.)

It should be added however, that apart from all of the moral problems like abortion surrounding this 'genetic counseling', whilst it undoubtedly does involve some direction of the course of evolution, it is a slow and not very efficient method of direction. To take a simple example: suppose one has an ailment caused by a recessive gene (*i.e.* just the homozygotes manifest the ailment). Suppose, just to keep the arithmetic simple, the gene's frequency is $1/100$ (*i.e.* in Hardy–Weinberg equilibrium, 1 person in 10,000 has the ailment). Even if one eliminated all of the people actually with the ailment (*i.e.* all the homozygotes), and even if one assumed that there was no mutation to the gene, in 100 generations the frequency would only drop to $1/200$ (*i.e.* be halved).

Moreover, one should be warned that this could all turn out to be a pretty expensive process and one may feel that there are more efficient and profitable ways of expending our resources and efforts in increasing the sum of human happiness. For instance, Tay-Sachs disease, something which proves fatal to small children and which is caused by having a certain recessive gene homozygously, affects between 45 and 50 people per year in the U.S. When one considers the cost of finding and eliminating the carriers, one may well think that one's time and money would be better spent in conventional forms of health care – think, for example, of the number of children today in North America that suffer mental retardation through malnutrition. (Hilton, 1973; Ruse, 1978a.)

Although there are yet other problems with genetic counseling (for example, what does one do about people who adamantly refuse to have diseased foetuses aborted, and what does one do about the offspring?), I am not arguing categorically against it. Nor am I arguing that other ways of directing

human biological evolution should not be explored. What I am arguing is that
it will not be a panacea for all of human ills and that the options must be left
open for other ways of increasing the general good. For example, if one can
find some simple environmental way of countering a genetic problem, why
not take it? (One thinks here of something like pyloric stenosis, a genetic
ailment involving blockage from the stomach, curable by a relatively simple
operation.)

9.6. SOCIOBIOLOGY AND THE DIRECTION OF EVOLUTION

We see, therefore, that already today human beings are starting to control and
to direct their biological evolution, although there is little doubt that the
uncontrolled evolution is a far greater factor. What does this all have to do
with sociobiology, or more pertinently, what does sociobiology have to do
with all of this? Obviously, inasmuch as any human social behaviour can be
shown to be a direct function of the genes, then it, like any other phenotypic
characteristic, becomes a candidate for possible manipulation or alteration or
elimination, as one attempts to direct the course of evolution. For example,
if one found that some grossly anti-social form of behaviour were caused by
the genes, then one might try to eliminate it by eliminating the carriers, as
we are now trying to eliminate Tay-Sachs disease. However, here even more
than elsewhere, I feel that the key to all of our future happiness does not lie
solely in the manipulation of future biological evolution.

First, there are going to be grave problems in deciding what constitutes
anti-social behaviour so gross that it ought to be eliminated, especially if this
involves elimination of the carriers. Homosexuality, if indeed it proves to
have a genetic base, would probably be the first and an on-going test case
here. One can well imagine the appalling social tensions that would be created
were any significant number of people to start advocating seriously the elimination of homosexual behaviour through amniocentesis screening programmes
and foetal abortion.

Second, even if certain characteristics, perhaps less than desirable, were
found to be genetic, it might well prove unfeasible or even counterproductive
to do anything about them. Suppose, for example, xenophobia (fear and
hostility towards strangers and outsiders) proved genetic, as Wilson (1975a)
sometimes suggests. *Prima facie*, it might seem a good idea to eliminate it:
think of the horrors of wars, and racial and religious prejudice, and the like.
But eliminating it would undoubtedly have to start with mass slaughter or
sterilization of the innocent. In a sense, Hitler's holocaust would be mild

beside it. Foetally speaking, I should think England would be virtually de-populated. Moreover, one might well find that as one threw out xenophobia, some very good human traits went too: perhaps, abstracting from the English, an ability to muddle through and stick things out when the going gets rough is pleiotropically linked to xenophobia. In other words, the attempt to design an ideal human society, genetically speaking, might prove prohibitively expensive and unattainable anyway. (I hasten to add that one can be an ardent sociobiologist without believing that one can and should attain such a perfect gene pool. Wilson himself points out the difficulties, if not impossibilities.)

Third, and most importantly, if we want to do something about human social behaviour, the most obvious place to start is with manipulation of the environment, not the genes. Suppose, for example, something like xenophobia were found to be controlled in some respects by the genes. It is hard to imagine that a systematic programme of education would have no effect on young people, even if the older people found it hard to change their ways. Consider, for instance, the case of anti-semitism in Germany. Without pretending that it is entirely gone — or elsewhere for that matter — it cannot be denied that it is far less in 1978 than it was in 1938. Obviously, this is a function of education and general social changes, rather than of changes in gene-ratios.

I am not denying that in some cases we might find some genetically based, social behavioural characteristics so destructive and so rigid (i.e. impervious to environmental manipulations) that we might feel that the only reasonable course of action is to try to prevent the existence of people with such characteristic-forming genes. But inasmuch as we develop grandiose plans for re-structuring human society, the obvious place to begin is with the environment, not the genes.

All in all therefore, the alteration of society through the alteration of genes and gene-ratios seems to be something more for the future and then perhaps of only limited value. But, in bringing to an end this part of the discussion, it must be pointed out that deliberately the discussion has been limited in scope, and that unless one recognizes this fact one may come away with an underestimation of the importance of sociobiology for future human happiness (or whatever else it is that one ought to maximize to achieve the greater good). Human evolution today is both biological and cultural, glossing over for the moment the extent to which the latter is a function of the former. Furthermore, it is the cultural side of humans which can change far more rapidly and which gives humans the greatest freedom of choice, allowing them in certain respects to escape from their biology. Consequently, as has just been pointed out, inasmuch as we are going to improve society, hoping thereby to increase human goods, it is undoubtedly culture which will prove to be the important sphere of action rather than biology.

Nevertheless, our understanding of human biology will be absolutely crucial to our cultural moves, for it will set limits to and show direction for these very moves. Consider a deliberately hypothetical example: suppose that we find that a certain recessive gene is very significantly correlated with a certain kind of repulsive anti-social behaviour; that, in fact, almost all who are homozygotes for this gene exhibit this undesirable behaviour. One way that one might try to eliminate the behaviour would be the purely biological one of detecting and aborting all homozygote foetuses. However, for various reasons, theoretical and practical, such a course of action may not be an open option. Consequently, alternatively one tries to eliminate the behaviour by taking the cultural route, that is by manipulating the environments of the affected homozygotes. But clearly, an awareness of the genetic origin of this behaviour could affect crucially the various plans of action that one tried. For instance, one might try various kinds of drug or diet therapy first, rather than be seeking to modify the familial environment by making the parents do, or not do, certain things. In other words, generalizing from this example, a knowledge of sociobiology could be absolutely vital as we try in the future to improve human social relations.

9.7. CONCLUSION

As I come to the end of this book, let me re-emphasize (in case the critics of sociobiology have forgotten) that in this and the last chapter I have deliberately assumed, for the sake of discussion, that the sociobiology of humans is a viable and fruitful enterprise. That I have made this assumption does not imply an unqualified endorsement. As I pointed out earlier, I am far from convinced that human sociobiologists have yet made their case. What I do plead is that their sins are not as grave as their critics argue. Human sociobiology should be given the chance to prove its worth. If it cannot deliver on its promises, it will collapse soon enough (Hull, 1978); but if it does prove viable, then its success could pay scientific dividends of the highest order.

BIBLIOGRAPHY

Adams, M.S., and J.V. Neel, (1967), 'Children of incest', *Pediatrics*, **40**, 55–62.
Achinstein, P., (1968), *Concepts of Science*, The Johns Hopkins Press, Baltimore.
Alexander, R.D., (1971), 'The search for an evolutionary philosophy', *Proc. Roy. Soc. Victoria Australia*, **84**, 99–120.
Alexander, R.D., (1974), 'The evolution of social behavior', *Ann. Rev. Ecology and Systematics*, **5**, 325–84.
Alexander, R.D., (1975), 'The search for a general theory of behavior', *Behavioral Science*, **20**, 77–100.
Alexander, R.D., (1977a), 'Evolution, human behaviour, and determinism', *PSA 1976*, F. Suppe and P. Asquith (Eds.), PSA, Michigan, pp. 3–21.
Alexander, R.D., (1977b), 'Natural selection and the analysis of human sociality'. In C.E. Goulden (Ed.), *Changing Scenes in Natural Sciences*, Philadelphia Academy of Natural Sciences, Philadelphia.
Alexander, R.D., and P.W. Sherman, (1977), 'Local mate competition and parental investment patterns in the social insects', *Science*, **196**, 494–50.
Allen, E., *et al.*, (1975), Letter to the Editor, *New York Review of Books*, **22**, 18, 43–4.
Allen, E., *et al.*, (1976), 'Sociobiology: another biological determinism', *BioScience*, **26**, 182–86.
Allen, E., *et al.*, (1977), 'Sociobiology: a new biological determinism'. In Sociobiology Study Group of Boston (Eds.), *Biology as a Social Weapon*, Burgess, Minneapolis.
Ayala, F.J., (1970), 'Teleological explanations in evolutionary biology', *Phil. Sci.*, **37**, 1–15.
Ayala, F., and Th. Dobzhansky, (1974), *Studies in the Philosophy of Biology*, University of California Press, California.
Ayala, F.J., M.L. Tracey, L.G. Barr, J.F. McDonald, and S. Perez-Salas, (1974), 'Genetic variation in natural populations of five Drosophila species and the hypothesis of the selective neutrality of protein polymorphisms', *Genetics*, **77**, 343–84.
Ayer, A.J., (1946), *Language, Truth and Logic*, 2nd edn., Gollancz, London.
Baerends, G.P., *et al.*, (1976), 'Multiple review of Wilson's *Sociobiology*', *Animal Behavior*, **24**, 698–718.
Barash, D.P., (1976), 'Some evolutionary aspects of parental behavior in animals and man', *Am. J. Psychology*, **89**, 195–217.
Barash, D.P., (1977), *Sociobiology and Behavior*, Elsevier, New York.
Bitterman, M., (1965), 'Phyletic differences in learning', *Am. Psychol.*, **20**, 396–410. In Seligman and Hager (Eds.), *Biological Boundaries of Learning*, Appleton-Century-Crofts, New York.
Block, N., and G. Dworkin, (1974), 'IQ, heritability and inequality', *Philosophy and Public Affairs*, **3**, 331–409; **4**, 40–99.
Bodmer, W., and L. Cavalli-Sforza, (1976), *Genetics, Evolution, and Man*, Freeman, San Francisco.

Bowler, P., (1976a), 'Malthus, Darwin, and the concept of struggle', *J. Hist. Ideas*, 37, 631–50.

Bowler, P., (1976b), *Fossils and Progress*, Science History Publications, New York.

Braithwaite, R.B., (1953), *Scientific Explanation*, Cambridge University Press, Cambridge.

Bunge, M., (1967), *Scientific Research*, Springer-Verlag, New York.

Bunge, M., (1968), 'Analogy in quantum theory: From insight to nonsense', *Brit. J. Phil. Sci.*, 18, 265–86.

Burchfield, J.D., (1975), *Lord Kelvin and the Age of the Earth*, Science History Publications, New York.

Burt, C., (1966), 'The genetic determination of differences in intelligence: A study of monozygotic twins reared together and apart', *Brit. J. Psychol.*, 57, 137–53.

Campbell, B., (1972), *Sexual Selection and the Descent of Man*, Aldine, Chicago.

Campbell, D.T., (1975), 'On the conflicts between biological and social evolution and between psychology and moral tradition', *American Psychologist*, 30, 1103–26.

Cancro, R., (1976), 'Genetic and environmental variables in schizophrenia'. In A. Kaplan (Ed.), *Human Behavior Genetics*, Thomas, Springfield, pp. 317–29.

Currier, R., (1976), 'Those beastly human genes' *Human Behavior*, pp. 16–22.

Darden, L., and N. Maull, (1977), 'Interfield theories', *Phil. Sci.*, 44, 43–64.

Dart, R.A., (1953), 'The predatory implemental technique of *Australopithecus*', *Anthropological and Linguistic Review*, 1, (4), 201–13.

Darwin, C., (1859), *On the Origin of Species by Means of Natural Selection*, Murray, London.

Darwin, C., (1871), *Descent of Man*, Murray, London.

Darwin, F., (1887), *The Life and Letters of Charles Darwin, Including an Autobiographical Chapter*, Murray, London.

Darwin, F., and A.C. Seward, (1903), *More Letters of Charles Darwin*, Murray, London.

Davie, M.R., (1929), *The Evolution of War*, Yale University Press, New Haven.

Dawkins, R., (1976), *The Selfish Gene*, Oxford University Press, Oxford.

de Beer, G., (1963), *Charles Darwin: Evolution by Natural Selection*, Nelson, London.

Dobzhansky, Th., (1937), *Genetics and the Origin of Species*. 3rd edn., 1951, Columbia University Press, New York.

Dobzhansky, Th., (1962), *Mankind Evolving*, Yale University Press, New Haven.

Dobzhansky, Th., (1970), *Genetics of the Evolutionary Process*, Columbia, New York.

Dobzhansky, Th., *et al.*, (1977), *Evolution*, Freeman, San Francisco.

Duhem, P., (1914), *The Aim and Structure of Physical Theory*. Princeton University Press, Princeton.

Durham, W.H., (1976a), 'The adaptive significance of cultural behavior', *Human Ecology*, 4, 89–121.

Durham, W.H., (1976b), 'Resource competition and human aggression, Part I: A review of primitive war', *Quart. Rev. Bio.*, 51, 385–415.

Durham, W.H., (1977), 'Reply to comments on "The adaptive significance of cultural behavior," ' *Human Ecology*, 5, 59–68.

Durham, W.H., (1978), 'The coevolution of human biology and culture', ms.

Durkheim, E., (1938), *The Rules of Sociological Method*, 8th edn., Free Press, New York.

Emlen, S.T., (1970), 'Celestial rotation: its importance in the development of migratory orientation', *Science*, 170, 1198–1201.

Ewing, A.C., (1953), *Ethics*, English Universities Press, London.
Feder, H.M., (1966), 'Cleaning symbioses in the marine environment'. In S.M. Henry (Ed.), *Symbiosis*, 1, Academic Press, New York, pp. 327–80.
Feyerabend, P., (1970), 'Against method: Outline of an anarchistic theory of knowledge'. In M. Radner and S. Winokur (Eds.), *Minnesota Studies in the Philosophy of Science*, Vol. 4, University of Minnesota Press, Minneapolis, pp. 17–130.
Fisher, R.A., (1930), *The Genetical Theory of Natural Selection*, Revised and reprinted 1958, Dover, New York.
Fisher, R.A., (1936), 'Has Mendel's work been rediscovered?' *Annals of Science*, 1, 115–137. Reprinted in *The Origins of Genetics: A Mendel Source Book*, C. Stern and E.R. Sherwood (Eds.), Freeman, San Francisco, 1966, pp. 139–72.
Flew, A.G.N., (1967), *Evolutionary Ethics*, Macmillan, London.
Fox, R., (1971), 'The cultural animal'. In J.F. Eisenberg and W.S. Dillon (Eds.), *Man and Beast: Comparative Social Behaviour*, Smithsonian Institution Press, Washington.
Freud, S., (1905), *Three Essays on the Theory of Sexuality*. In *Collected Works of Freud* J. Strachey (Ed.), Vol. 7, Hogarth, London, 1953.
Freud, S., (1913), *Totem and Taboo*. In J. Strachey (Ed.), *Collected Works of Freud* Vol. 13, Hogarth, London, 1953.
Garcia, J., F.R. Ervin, and R. Koelling, (1966), 'Learning with prolonged delay of reinforcement', *Psychonomic Science*, 5, 121–2.
Garcia, J., and R. Koelling, (1966), 'Relation of cue to consequence in avoidance learning', *Psychonomic Science*, 4, 123–4.
Garcia, J., B.K. McGowan, and K.F. Green, (1972), 'Biological constraints on conditioning'. In A.H. Black and W.F. Prokasy (Eds.), *Classical Conditioning II: Current Research and Theory*, Appleton-Century-Crofts, New York.
George, W., (1964), *Elementary Genetics*, 2nd edn., Macmillan, London.
Ghiselin, M., (1969), *The Triumph of the Darwinian Method*, University of California Press, Berkeley.
Ghiselin, M., (1974), *The Economy of Nature and the Evolution of Sex*, University of California Press, Berkeley.
Goudge, T.A., (1961), *The Ascent of Life*, University of Toronto Press, Toronto.
Haldane, J.B.S., (1955), 'Population genetics', *New Biology*, 18, 34–51.
Hamilton, W.D., (1964a), 'The genetical theory of social behaviour. I', *J. Theor. Biol.*, 7, 1–16.
Hamilton, W.D., (1964b), 'The genetical theory of social behaviour. II', *J. Theor. Biol.*, 7, 17–32.
Hamilton, W.D., (1967), 'Extraordinary sex ratios', *Science*, 156, 477–88.
Hamilton, W.D., (1971), 'Geometry for the selfish herd', *J. Theor. Biol.*, 31, 295–311.
Hamilton, W.D., (1972), 'Altruism and related phenomena, mainly in social insects', *Ann. Rev. Ecol. Syst.*, 3, 193–232.
Harris, M., (1971), *Culture, Man, and Nature: An Introduction to General Anthropology*, Crowell, New York.
Hartung, J., (1976), 'On natural selection and the inheritance of wealth', *Current Anthropology*, 17, 606–14.
Hempel, C.G., (1965), *Aspects of Scientific Explanation*, Free Press, New York.
Hempel, C.G., (1966), *Philosophy of Natural Science*, Prentice-Hall, Englewood Cliffs.
Herrnstein, R.J., (1971), 'IQ', *Atlantic Monthly*, September, 43–64.

Herrnstein, R.J., (1975), 'Kamin errs . . . Herrnstein', *Contemp. Psychol.*, **20**, 758.
Hilton, B., *et al.*, (1973), *Ethical Issues in Human Genetics*, Plenum, New York.
Himmelfarb, G., (1968), 'Varieties of social Darwinism'. In *Victorian Minds*, Weidenfeld and Nicolson, London.
Hinde, R.A., (1970), *Animal Behaviour: A Synthesis of Ethology and Comparative Psychology*, McGraw-Hill, New York.
Hudson, W.D., (1970), *Modern Moral Philosophy*, Macmillan, London.
Hull, D.L., (1972), 'Reduction in genetics – biology or philosophy?' *Phil. Sci.*, **39**, 491–99.
Hull, D.L., (1973), 'Reduction in genetics – doing the impossible'. In P. Suppes *et al.* (Eds.), *Logic, Methodology and Philosophy of Science*, **IV**, 619–35.
Hull, D.L., (1973), *Darwin and His Critics*, Harvard University Press, Cambridge, Mass.
Hull, D.L., (1974), *Philosophy of Biological Science*, Prentice-Hall, Englewood Cliffs.
Hull, D.L., (1976), 'Informal aspects of theory reduction'. In A.C. Michalos *et al.* (Eds.), *PSA 1974*, Reidel, Dordrecht.
Hull, D.L., (1977), 'A logical empiricist looks at biology', *Brit. J. Phil. Sci.*, **28**, 181–9.
Hull, D.L., (1978a), 'Sociobiology: Scientific bandwagon or traveling medicine show? *Transaction Society* **15**(6), 50–9.
Hull, D.L., (1978b), 'The trouble with traits', *Theory and Decision*. Forthcoming.
Hume, D., (1740), *Treatise of Human Nature*, London.
Huxley, J.S., (1938), 'The present standing of the theory of sexual selection'. In G. de Beer (Ed.), *Evolution*, Clarendon Press, Oxford. pp. 11–42.
Huxley, J., (1947), *Evolution and Ethics*, Pilot, London.
Huxley, T.H., (1893), 'Evolution and ethics' (The Romanes Lecture for 1893). In *Collected Essays*, 9. Macmillan, London.
Jencks, C., *et al.*, (1972), *Inequality*. Basic Books, New York.
Jensen, A.R., (1969), 'How much can we boost IQ and scholastic achievement?' *Harvard Educational Review*, **39**, 1–123.
Jensen, A., (1972), *Genetics and Education*, Harper and Row, New York.
Jonas, H., (1976), 'Freedom of scientific inquiry and the public interest', *Hastings Center Report*, August, 1976, pp. 15–17.
Kallman, F.J., (1952), 'Twin and Sibship study of overt male homosexuality', *Am. J. Hum. Gen.*, **4**(2), 136–46.
Kamin, L., (1974), *Science and Politics of IQ*, Erlbaum, Potomic.
Kamin, L., (1977a), 'Comment on Munsinger's adoption study', *Behav. Gen.*, **7**, 403–6.
Kamin, L., (1977b), 'A reply to Munsinger', *Behav. Gen.*, **7**, 411–12.
Kleiner, S., (1975), 'Essay review: the philosophy of biology', *Southern J. of Phil.*, **13**, 523–42.
Kuhn, T.S., (1970), *The Structure of Scientific Revolutions*, 2nd edn., Chicago University Press, Chicago.
Lack, D., (1947), *Darwin's Finches: An Essay on the General Biological Theory of Evolution*, Cambridge University Press, Cambridge.
Lack, D., (1954), *The Natural Regulation of Animal Numbers*, Oxford University Press, Oxford.
Lack, D., (1966), *Population Studies of Birds*, Oxford University Press, Oxford.
Leibenstein, H., (1976), *Beyond Economic Man: A New Foundation for Economics*, Harvard University Press, Cambridge, Mass.

Levins, R., (1968), *Evolution in Changing Environments: Some Theoretical Explorations*, Princeton University Press, Princeton.

Levins, R., (1970), 'Extinction'. In M. Gerstenhaber, (Ed.), *Some Mathematical Questions in Biology*, American Mathematical Society, Providence, pp. 77–107.

Lewontin, R.C., (1961), 'Evolution and the theory of games', *J. Theor. Biol.*, **1**, 382–403.

Lewontin, R.C., (1970), 'The Units of Selection', *Annual Review of Ecology and Systematics*, **1**, R.F. Johnston *et al.* (Eds.), Annual Reviews Inc., California.

Lewontin, R.C., (1972), 'The apportionment of human diversity', *Evolutionary Biology*, **6**, 381–98.

Lewontin, R.C., (1974), *The Genetic Basis of Evolutionary Change*, Columbia University Press, New York.

Lewontin, R.C., (1977), 'Sociobiology – A caricature of Darwinism'. In F. Suppe and P. Asquith (Eds.), *PSA 1976*, Vol. 2, PSA, Lansing, Mich.

Lewontin, R.C., and M.W. Feldman, (1975), 'The heritability hang-up', *Science*, **190**, 1163–8.

Li, C.C., (1955), *Population Genetics*, Chicago University Press, Chicago.

Livingstone, F.B., (1967), *Abnormal Hemoglobins in Human Populations*, Aldine, Chicago.

Livingstone, F.B., (1971), 'Malaria and human polymorphisms', *Annual Review of Genetics*, **5**, 33–64.

Locke, J., (1959), *An Essay Concerning Human Understanding*, A.C. Fraser (Ed.), Dover, New York.

Lorenz, K., (1966), *On Aggression*, Harcourt Brace and World, New York.

MacArthur, R.H., and E.O. Wilson, (1967), *The Theory of Island Biogeography*, Princeton University Press, Princeton.

Mansfield, E., (1970), *Microeconomic: Theory and Applications*, Norton, New York.

Marmor, J., (1965), *Sexual Inversion: The Multiple Roots of Homosexuality*, Basic Books, New York.

Maynard, E.C.L., (1968), 'Cleaning symbiosis and oral grooming on the coral reef'. In P. Person (Ed.), *Biology of the Mouth*, AAAS, Wash. D.C., pp. 79–88.

Maynard Smith, J., (1972), 'Game theory and the evolution of fighting'. In *On Evolution*, Edinburgh University Press, Edinburgh.

Maynard Smith, J., (1974), 'The theory of games and the evolution of animal conflict', *J. Theor. Biol.*, **47**, 209–21.

Maynard Smith, J., (1975), *The Theory of Evolution*, 3rd edn., Penguin, Harmondsworth.

Maynard Smith, J., (1976), 'Evolution and the theory of games', *Amer. Sci.*, **64**, 41–5.

Maynard Smith, J., and G. Parker, (1976), 'The logic of asymmetric contests', *Anim. Behav.*, **24**, 159–75.

Maynard Smith, J., and G. Price, (1973), 'The logic of animal conflicts', *Nature*, **246**, 15–18.

Mayr, E., (1942), *Systematics and the Origin of Species*, Columbia University Press, New York.

Mayr, E., (1963), *Animal Species and Evolution*, Belknap, Cambridge, Mass.

Mayr, E., (1969), 'Scientific explanation and conceptual framework', *J. Hist. Biol.*, **2**, 123–8.

Mayr, E., (1974), 'Teleological and teleonomic, a new analysis'. In R.S. Cohen and M.W.

Wartofsky (Eds.), *Boston Studies in the Philosophy of Science*, **XIV**. Reidel, Boston, pp. 91–117.

McClearn, G.E., and J.C. DeFries, (1973), *Introduction to Behavioral Genetics*, Freeman, San Francisco.

Mettler, L.E., and T.G. Gregg, (1969), *Population Genetics and Evolution*, Prentice-Hall, Englewood Cliffs.

Moore, G.E., (1903), *Principia Ethica*, Cambridge University Press, Cambridge.

Munsinger, H., (1975a), 'Children's resemblance to their biological and adopting parents in two ethnic groups', *Behav. Gen.*, **5**, 239–54.

Munsinger, H., (1975b), 'The adopted child's IQ: A critical review', *Psychol. Bull.*, 82, 623–59.

Munsinger, H., (1977), 'A reply to Kamin', *Behav. Gen.*, **7**, 407–9.

Munson, R., (1971), *Man and Nature: Philosophical Issues in Biology*, Delta, New York.

Murdock, G.P., (1967), *Ethnographic Atlas*, University of Pittsburgh Press, Pittsburgh.

Murphy, R.F., (1957), 'Intergroup hostility and social cohesion', *Am. Anthropol.*, **59**, 1018–35.

Nagel, E., (1961), *The Structure of Science*, Routledge and Kegan Paul, London.

Neill, A.S., (1960), *Summerhill: A Radical Approach to Child Rearing*, Hart, New York.

Ospovat, D., (1976), 'The influence of Karl Ernst von Baer's embryology, 1828-1859: A reappraisal in light of Richard Owen's and William B. Carpenter's "Palaeontological application of 'von Baer's law' " ', *J. Hist. Biol.*, **9**, 1–28.

Oster, G.F., and E.O. Wilson, (1978), *Caste and Ecology in the Social Insects*, Princeton University Press, Princeton.

Owen, R., (1848), *On the Archetype and Homologies of the Vertebrate Skeleton*, Voorst, London.

Pare, C., (1965), 'Etiology of homosexuality: genetic and chromosomal aspects'. In J. Marmor (Ed.), *Sexual Inversion: The Multiple Roots of Homosexuality*, Basic Books, New York.

Parker, S., (1976), 'The precultural basis of the incest taboo: toward a biosocial theory', *American Anthropologist*, 78, 285–305.

Peel, J.D.Y., (1971), *Herbert Spencer: The Evolution of a Sociologist*, Heinemann, London.

Popper, K.R., (1959), *The Logic of Scientific Discovery*, Hutchinson, London.

Provine, W.B., (1971), *The Origins of Theoretical Population Genetics*, Chicago University Press, Chicago.

Quinton, A., (1966), 'Ethics and the theory of evolution'. In I.T. Ramsey (Ed.), *Biology and Personality*, Blackwell, Oxford.

Rainer, J.D., (1976), 'Genetics and homosexuality'. In A. Kaplan (Ed.), *Human Behavior Genetics*, Thomas, Springfield.

Raper, A.B., (1960), 'Sickling and malaria', *Trans. Roy. Soc. Trop. Med. Hyg.*, **54**, 503–4.

Raphael, D.D., (1958), 'Darwinism and ethics'. In S.A. Barnett (Ed.), *A Century of Darwin*, Heinemann, London.

Rudwick, M.J.S., (1964), 'The inference of function from structure in fossils', *Brit. J. Phil. Sci.*, **15**, 27–40.

Ruse, M., (1969), 'Confirmation and falsification of theories of evolution', *Scientia*, civ, 329–57.

Ruse, M., (1970), 'The revolution in biology', *Theoria,* **XXXVI,** 1–22.

Ruse, M., (1971a), 'Two biological revolutions', *Dialectica,* **25,** 17–38.

Ruse, M., (1971b), 'Natural Selection in "The Origin of Species" ', *Studies in the History and Philosophy of Science,* **1,** 311–351.

Ruse, M., (1972), 'Is the theory of evolution different?', *Scientia,* **106,** 765–83; 1069–93.

Ruse, M., (1973a), *The Philosophy of Biology,* Hutchinson, London.

Ruse, M., (1973b), 'The value of analogical models in science', *Dialogue,* **XII,** 246–53.

Ruse, M., (1973c), 'The nature of scientific models: formal v material analogy', *Phil. Soc. Sci.,* **3,** 63–80.

Ruse, M., (1974), 'Cultural evolution', *Theory and Decision,* **5,** 413–40.

Ruse, M., (1975a), 'Darwin's debt to philosophy: an examination of the influence of the philosophical ideas of John F.W. Herschel and William Whewell on the development of Charles Darwin's theory of evolution', *Stud. Hist. Phil. Sci.,* **6,** 159–81.

Ruse, M., (1975b), 'The relationship between science and religion in Britain, 1830–1870', *Church History,* **44,** 505–22.

Ruse, M., (1975c), 'Charles Darwin and artificial selection', *J. Hist. Ideas,* **36,** 339–50.

Ruse, M., (1975d), 'Charles Darwin's theory of evolution: an analysis', *J. Hist. Biol.,* **8,** 219–41.

Ruse, M., (1976a), 'Reduction in genetics'. In R.S. Cohen *et al.* (Eds.), *PSA 1974,* Reidel Dordrecht, pp. 633–52.

Ruse. M., (1977a), 'Is biology different from physics?'. In R. Colodny (Ed.), *Logic, Laws, and Life,* University of Pittsburgh Press, Pittsburgh, pp. 89–127.

Ruse, M., (1977b), 'William Whewell and the argument from design', *Monist,* **60,** 244–68.

Ruse, M., (1977c), 'Sociobiology: Sound science or muddled metaphysics?'. In F. Suppe and P. Asquith (Eds.), *PSA 1976,* Vol. 2, pp. 48–73.

Ruse, M., (1977d), 'Karl Popper's philosophy of biology', *Philosophy of Science,* **44,** 638–61.

Ruse, M., (1978a), 'Human genetic technology: the darker side'. In J.E. Thomas (Ed.), *Matters of Life and Death,* Hackett, Toronto.

Ruse, M., (1978b), 'Philosophical factors in the Darwinian Revolution'. In F. Wilson (Ed.), *Pragmatism and Purpose,* University of Toronto Press, Toronto.

Ruse, M., (1978c), 'The dangers of unrestricted research: the case of recombinant DNA'. In J. Richards (Ed.), *Science, Ethics, and Politics: The Recombinant DNA Debate,* Harper, New York.

Sahlins, M.D., (1965), 'On the sociology of primitive exchange'. In M. Banton (Ed.), *The Relevance of Models for Social Anthropology,* Tavistock, London, pp. 139–236.

Sahlins, M.D., (1976), *The Use and Abuse of Biology,* University of Michigan, Ann Arbor.

Sahlins, M.D., and E. Service, (1960), *Evolution and Culture,* University of Michigan Press, Ann Arbor.

Salmon, W., (1973), *Logic,* Prentice-Hall, Englewood Cliffs.

Schaffner, K.F., (1967), 'Antireductionism and molecular biology', *Science,* **157,** 644–7.

Schaffner, K.F., (1969), 'The Watson–Crick model and reductionism', *Brit. J. Phil. Sci.,* **20,** 325–48.

Schilpp, P., (1974), *The Philosophy of Karl Popper.* (Ed.), Open Court, La Salle, Ill.

Seligman, M., (1971), 'Phobias and preparedness', *Behavior Therapy,* **2,** 307–20.

Seligman, M., and J. Hager, (1972), *Biological Boundaries of Learning*, Appleton-Century-Crofts, New York.

Sheppard, P.M., (1975), *Natural Selection and Heredity*, 4th edn., Hutchinson, London.

Shields, J., (1962), *Monozygotic Twins Brought Up Apart and Brought Up Together*, Oxford University Press, London.

Siegel, S., (1956), *Nonparametric Statistics for the Behavioral Sciences*, McGraw-Hill, New York.

Simpson, G.G., (1944), *Tempo and Mode in Evolution*, Columbia University Press, New York.

Simpson, G.G., (1953), *The Major Features of Evolution*, Columbia University Press, New York.

Spencer, H., (1850), *Social Statics*, Chapman, London.

Spencer, H., (1852a), 'The development hypothesis', *Leader*. Reprinted in *Essays*, 1, 377–83.

Spencer, H., (1852b), 'A theory of population, deduced from the general law of animal fertility', *Westminster Review*, n.s. 1, 468–501.

Spencer, H., (1857), 'Progress: its law and cause', *Westminster Review*. In *Essays*, 1, 1–60.

Stevenson, C.L., (1944), *Ethics and Language*, Yale University Press, New Haven.

Strickberger, M.W., (1968), *Genetics*, Macmillan, New York.

Suppe, F., (1974), *The Structure of Scientific Theories*, University of Illinois Press, Urbana.

Tinbergen, N., (1951), *The Study of Instinct*, Clarendon Press, Oxford.

Tinbergen, N., (1953), *Social Behaviour in Animals*, Methuen, London.

Townes, B.D., W.D. Ferguson, and S. Gillam, (1976), 'Differences in psychological sex adjustment and familial influences among homosexual and non-homosexual populations', *J. Homosexuality*, 1, 261–72.

Trivers, R.L., (1971), 'The evolution of reciprocal altruism', *Quart. Rev. Biol.*, 46, 35–57.

Trivers, R.L., (1972), 'Parental investment and sexual selection'. In B. Campbell (Ed.), *Sexual Selection and the Descent of Man, 1871–1971*, Aldine, Chicago.

Trivers, R.L., (1974), 'Parent-offspring conflict', *Am. Zoo.*, 14, 249–64.

Trivers, R.L., (1976), 'Foreword' to R. Dawkins, *The Selfish Gene*, Oxford University Press, Oxford, pp. v–vii.

Trivers, R.L., and H. Hare., (1976), 'Haplodiploidy and the evolution of social insects', *Science*, 191, 249–63.

Trivers, R.L., and D.E. Willard, (1973), 'Natural selection of parental ability to vary the sex ratio of offspring', *Science*, 179, 90–92.

Vandenberg, S.G., (1976), 'Twin studies'. In A. Kaplan (Ed.), *Human Behavior Genetics*, Thomas, Springfield.

Van den Berghe, P.L., and D.P. Barash, (1977), 'Inclusive fitness and human family structure', *American Anthropologist*, 79, 809–823.

Vorzimmer, P.J., (1970), *Charles Darwin: The Years of Controversy*, Temple University Press, Philadelphia.

Waddington, C.H., (1960), *The Ethical Animal*, Allen and Unwin, London.

Wade, N., (1976a), 'Sociobiology: troubled birth for new discipline', *Science*, 191, 1151–5.

Wade, N., (1976b), 'IQ and heredity: Suspicion of fraud beclouds classic experiment', *Science*, 194, 916–19.

Wallace, A.R., (1870), 'A theory of bird's nests'. In *On Natural Selection*, Macmillan, London, pp. 231–61.

Walsh, J., (1976), 'Science for the people: comes the revolution', *Science*, 191, 1033–5.

Warnock, G.J., (1976), *Contemporary Moral Philosophy*, Macmillan, London.

Watson, J.D., (1968), *The Double Helix*, Atheneum, New York.

Watson, J.D., (1970), *Molecular Biology of the Gene*, 2nd edn., Benjamin, New York.

West Eberhard, M.J., (1975), 'The evolution of social behavior by kin selection', *Quart. Rev. Biol.*, 50, 1–33.

White, L., (1959), *The Evolution of Culture*, McGraw-Hill, New York.

Williams, G.C., (1966), *Adaptation and Natural Selection: A Critique of Some Current Evolutionary Thought*, Princeton University Press, Princeton.

Williams, G.C., (1975), *Sex and Evolution*, Princeton University Press, New Jersey.

Williams, M.B., (1970), 'Deducing the consequences of evolution: a mathematical model', *J. Theor. Biol.*, 29, 343–385.

Wilson, E.O., (1971), *The Insect Societies*, Belknap, Cambridge, Mass.

Wilson, E.O., (1975a), *Sociobiology: The New Synthesis*, Belknap, Cambridge, Mass.

Wilson, E.O., (1975b), 'Human decency is animal', *The New York Times Magazine*, 12, October, 38–50.

Wilson, E.O., (1975c), 'Letter to the Editor', *New York Review of Books*, 22, 20, 60–1.

Wilson, E.O., (1976), 'Academic vigilantism and the political significance of sociobiology', *BioScience*, 26, 183–90.

Wilson, E.O., (1977a), 'Biology and the social sciences', *Daedalus*.

Wilson, E.O., (1977b), 'Foreword' to D.P. Barash, *Sociobiology and Behavior*, Elsevier, New York.

Wright, S., (1966), 'Mendel's Ratios'. In C. Stern and E.R. Sherwood (Eds.), *The Origins of Genetics: A Mendel Source Book*, Freeman, San Francisco, pp. 173–5.

Wynne-Edwards, V.C., (1962), *Animal Dispersion in Relation to Social Behaviour*, Oliver and Boyd, Edinburgh.

SUBJECT INDEX

99, 143, 145, 150, 153, 158, 186,
  187, 209
Dominance (genetic) 8
Down's Syndrome 92–3, 130, 133
Drosophila 103, 128–129

Economics 107, 126, 168, 189–91
Elephant seal 32
Embryology 7, 18, 115, 148, 164
Emotivism 205
Englishmen 78, 199, 213
Environment 8–9, 12, 23, 45, 74, 78,
  100, 107, 114, 129–33, 138–9, 141,
  154, 175, 180, 198, 213–14
Epigamic selection 38, 94
Epistemology 20
Equal Rights Amendment 100
Eskimo 59, 173, 175
Ethics, see Evolutionary ethics or
  Morality
Ethology 22, 25, 182–3
Eugenics 75
Evolution 6–7, 10–11, 16–21, 32–3,
  54–5, 60, 62, 69, 80–1, 84–5, 91,
  107, 109–10, 113–15, 148–50,
  156–7, 161, 184, 195, 199–202, 204,
  210, 212
Evolutionary ethics 85, 194–214
Evolutionary stable strategies 26–31, 110

Fallacy 104–6, 200–1
Falsifiability 106, 111–119
Falsification 105, 112, 119–25
Female 32–8, 45–6, 58–60, 62, 67, 86,
  89, 93–100, 124–5, 143, 145, 148–
  50, 153, 155, 157–8, 161, 187, 207
Female choice 32, 38, 57, 60
Fertilization 33, 35–6, 45
Fighting, see aggression
Fish 30, 35–6, 50, 95–6, 158
Fitness 11–13, 38, 44, 66, 91–2, 110,
  118, 155, 157, 186
Fossil record 6, 148
Founder principle 18
Free-enterprise system, see Capitalism
Frenchmen 78
Fruit-fly, see Drosophila

Galapagos 16
Game-theory 26–31, 37, 50, 56–7,
  146–8
Gamete 33
Gene, *passim*
Genetic counseling 211
Genetic drift 114–15
Genotype 8, 10, 64, 130, 204
Geospizinae, 17
Group selection 13–16, 23, 28, 31, 36–
  7, 42–3, 49, 54, 56, 107–10, 149,
  151, 196
Guilt 70–1

Haploid 9, 45
Hardy-Weinberg law 10, 179, 211
He-man strategy 36–8, 59–60, 96
Herring gull 182–3
Heterosexuality 90, 93, 155, 186, 208
Heterozygote 8, 12–13, 60, 66, 91–2,
  118, 186
Hindu taboos 177–9, 181
Homo sapiens 1, 12, 23, 52–76, 79, 81,
  83–9, 93–4, 96, 98, 108, 110, 116,
  118, 120, 122, 124, 127, 129, 131–2,
  134–5, 138, 141–6, 150–2, 157,
  161, 163, 165, 170, 185, 188, 190,
  194, 199, 202, 209, 211
Homosexuality 26, 66, 69, 85, 89, 90–3,
  118, 133–5, 141, 153–6, 186–9, 212
Homozygote 7, 10, 12, 66, 91, 211, 214
Human, see *Homo sapiens*
Hymenoptera 6, 24, 45–8, 104, 108, 117,
  124, 148–50, 191
Hypocrisy 70, 87, 89
Hypothetico-deductive system 20–1,
  181, 193

Imprinting 183
Inbreeding, see Incest
Incest 60–1, 119, 124, 132–4, 141, 153,
  156, 162–4, 178, 187, 188–9,
Inclusive fitness 41, 44, 66, 124
Indigo buntings 184–5
Individual selection 13–16, 25–8, 36,
  42–3, 49–50, 86, 109, 111, 116–17,
  149, 151–2, 197

NAME INDEX

Hempel, C. G. 106, 112, 146
Herrnstein, R. J. 136, 140
Hilton, B. 211
Himmelfarb, G. 53, 81, 199
Hinde, R. A. 9
Hitler, A. 75, 212
Hobbes, T. 100
Hudson, W. D. 87, 199, 201
Hull, D. L. 4, 8, 10, 18, 21, 108, 114,
  126, 164, 166, 168, 214
Hume, D. 200
Huxley, J. S. 38, 107, 112, 194
Huxley, T. H. 194

Imo 144, 149

Jencks, C. 138
Jensen, A. 136, 138
Joan of Arc 78, 132
Jonas, H. 76

Kallman, F. J. 133–34, 156
Kamin, L. 136–8
Kant, I. 201–3, 208
Kerkule 82
Kleiner, S. 166
Koelling, R. 184
Koestler, A. 82
Kropotkin, P. 199
Kuhn, T. S. 2–3, 19–21, 166

Lack, D. 15–17, 32, 109
Lamarck, J. B. 81
Leibenstein, H. 190–1
Levins, R. 2, 4, 109
Levi-Strauss, C. 178
Lewontin, R. C. 2, 4, 13, 79, 94, 102–3,
  109, 114, 119–20, 126, 128–9,
  167–8
Li, C. C. 10
Livingstone, F. B. 12
Locke, J. 164
Lorenz, K. 22–6, 54, 56, 75, 149, 182–3
Lyell, C. 81

MacArthur, R. H. 4
Malthus, T. R. 80–3, 100, 108

Mansfield, E. 189–90
Marmor, J. 133, 186
Maull, N. 166
Maynard, E. C. L. 50
Maynard Smith, J. 16, 26–31, 49, 53,
  56–7, 110, 146–8
Mayr, E. 10, 16, 18, 107–8, 168
McClearn, G. E. 131
McGowan, B. 184
Mendel, G. 9, 44–5, 105, 138, 167
Mettler, L. E. 10
Moore, G. E. 200
Morgan, T. H. 206
Morris, D. 3
Munsinger, H. 136–7
Munson, R. 195
Murdock, G. P. 123, 159
Murphy, R. F. 173

Nagel, E. 166
Neel, J. V. 163
Neill, A. S. 153

Ospovat, D. 164
Oster, G. F. 4, 47, 51, 117, 126, 191
Owen, R. 166

Pare, C. 133–4, 187
Parker, G. 29
Parker, S. 162
Peel, J. D. Y. 199
Plato 135
Popper, K. R. 106, 111–19
Price, G. 27–9
Provine, W. B. 108
Ptolemy, C. 206

Quinton, A. 194, 200

Rainer, J. D. 133
Raper, A. B. 12
Raphael, D. D. 194, 206
Rockefeller, J. D. 53, 79
Rudwick, M. J. S. 114
Ruse, M. 4, 7, 11, 18, 21–2, 76, 81, 114,
  127, 142, 161, 163, 166, 168, 210–11

EPISTEME

A SERIES IN THE FOUNDATIONAL, METHODOLOGICAL,
PHILOSOPHICAL, PSYCHOLOGICAL, SOCIOLOGICAL,
AND POLITICAL ASPECTS OF THE SCIENCES, PURE AND APPLIED

Editor: MARIO BUNGE
Foundations and Philosophy of Science Unit, McGill University